UNHERALDED
HEROES

by

Thomas van Hees

The contents of this book regarding the accuracy of events, people, and places depicted; permissions to use all previously published materials; opinions expressed; all are the sole responsibility of the author, who assumes all liability for the contents of this book and indemnifies the publisher against any claims stemming from publication of this book.

International Standard Book Number 13: 978-1-60452-080-4
International Standard Book Number 10: 1-60452-080-9

Library of Congress Control Number: 2013949237

BluewaterPress LLC
52 Tuscan Way Ste 202-309
Saint Augustine Florida 32092

http://bluewaterpress.com

This book may be purchased online at -
http://bluewaterpress.com/heroes

Acknowledgments

I would like to thank all of the heroes, past and present, in my life. My dad told me early on that if I wanted to succeed in life that I must work hard to provide for those who I love. Dad is my number one hero from the past. My present day hero is the love of my life, Judi, for whom without her I would have stumbled and fell long ago.

Dedication

"I was not a hero but I stood along the
side of men that were"

To all heroes that have molded our country's history and
for those who will continue to make history, we thank you.

Through His Eyes

As I crawled to my spot
Wishing I wouldn't have fought

I set up my scope
Full of hope

I hear the screams and shiver
The artillery raining down makes me quiver

I load my rifle
Put my thoughts to a stifle

I dial the scope's knob
Wanting a different job

Getting ready for the shot
Watching soldiers being caught

Took my aim
Without shame

I fired
Wishing he hadn't been hired

The wind blew silent
Then I ran

By
Cory Van Hees

Introduction

Webster's New World Dictionary defines hero as "any person, admired for courage, nobility, etc." In the thousands upon thousands of words in the English language, one word stands out as the most used and misused word that pinpoints the achievements of a single person. Hero is this word. A person can be tagged a hero, they can be labeled a hero and they can be called a hero. Some may seem to be a hero for actions they have not yet accomplished, when some have already become heroes and do not know it.

We live in a country that has been involved in war since its existence and will continue to be involved because this is what we, as keepers and believers of freedom, do. It will never change and as long as we are involved, we will continue to repeatedly produce heroes.

Make no mistake war will certainly make heroes, but war is not the only way to become a hero. Everyday life produces heroes just as well.

Does a single heroic feat earn this label or do all involved deserve this title? Although there is only one true definition of hero, many have their own interpretation of the meaning of the word.

Are heroes made or do they just become one? Who and what determines a hero? In this book I will try to explain who are the real heroes and show that heroes need not be brave to warrant being called a hero, they just need to succeed in life.

Table of Contents

Acknowledgments   vii

Dedication   ix

Introduction   xiii

Revolutionary War Heroes: Fact or Fiction   1

Daniel Morgan   2

Henry Lee III   3

Alexander Hamilton   3

William Alexander   3

George Washington   3

The Civil War   5

The 27th Maine Volunteer Infantry Regiment   5

Revoked Medal of Honor Awards   6

Unsung Heroes   7

Surprisingly Young Unsung Heroes   9

Women In The Civil War   12

Blacks In The Civil War   14

Photographers of The Civil War   16

Mathew B. Brady   16

Alexander Gardner   17

James F. Gibson   18

Southern Photographers   18

George S. Cook   18

Robert M. Smith   18

Surprising Unsung Heroes   19

Inanimate Heroes   20
James B. Eads   20
World War I   23
The WW I Heroes of Technology   24
Hugo Junker   25
Roland Garros   25
The Land-Ships of World War I   27
The Unsung Heroes of World War I   28
Harry L. Wingate   28
Black Soldiers In World War I   28
The Red Hand Division   29
Henry Lincoln Johnson   29
Eugene Bullard   30
The Code Talkers   32
The Merchant Marine Service   34
Women and The War   35
Nursing On The Front Lines   37
Keeping The Home Fires Burning   39
Animals In WWI   39
Sergeant Stubby   40
World War II   43
War Time Inventions   44
The Higgins Boat   45
Liberty Ships   45
Operation Pluto   47
The Ladies Of War   49
Rosie The Riveter   49
Martha Gellhorn   50
Martha Raye   51
Josephine Baker   54

Nancy Wake 55

Soviet Women In World War II 57

Women Pilots 57

Land Forces 58

Women Air Force Service Pilots (WASP) 59

Air Spotters of WW II 60

The Merchant Seaman of World War II 62

Women POW's of World War II 65

World War II Celebrities 66

Lee Marvin 66

Don Adams 66

Gene Autry 66

Eddie Albert 67

James Arness 67

Mel Brooks 68

Charles Bronson 68

Charles Durning 68

Art Carney 69

Rod Serling 69

James (Jimmy) Stewart 71

Ed McMahon 73

Charles Schulz 74

Julia Child 74

The Korean War 77

The C-Ration Meal 78

Medevac Helicopters 79

Mobile Army Surgical Hospital (MASH) 81

Engineer LeRoy Stelck 83

Jet Aircraft In Korea 85

Women In The Korean War 86

The Vietnam War   89
Vietnam War Photojournalists   89
Sean Flynn   90
Dickey Chapelle   91
Hugh Van Es   92
The Bird Dog   93
Gun Trucks: A Vietnam Innovation   95
Tunnel Rats   97
The U.S. Coast Guard In Vietnam   98
Women In Vietnam   99
The USO Club Women   100
Vietnamese Women At War   102
Women POW's In Vietnam   103
The following missionaries were POWs: 104
The Medevacs Of Vietnam   104
Medevac Pilots   107
Michael J. Novosel   107
Bruce P. Crandall   108
Ed "Too Tall" Freeman   108
Hugh Thompson Jr.   109
The Secret Hidden War   114
Honored But Not Forgotten   116
The Persian Gulf Wars   119
Precision Guided Munitions   119
Satellite Directed Systems   120
Airborne Warning and Control System (AWACS)   121
Photocopiers   121
The Patriot Missile   121
The Gun Truck: A Vietnam Innovation Returns   121
Women In The Gulf Wars   123

Leigh Ann Hester   125
Monica Lin Brown   125
Women POW's Of The Gulf Wars   126
Journalists In War   127
The True Unheralded Heroes   128
The Author's Perspective   131
Information References   132

Revolutionary War Heroes: Fact or Fiction

The Revolutionary War has been said to produce many heroes, but in my opinion, few were recognized as being one. A war that lasted for nearly eight years and involved more than 250,000 troops, of which only 90,000 were armed, there was only a small percentage that were labeled as heroes.

Military battle tactics were anything but strategically used because those who led were just inept or not trained enough to put a satisfactory battle plan in place. Those who were in charge had great ideas about how a battle should be performed. Due to the lack of experience, or battle plans that had not been tried, many situations failed miserably. One failure after another led to a better battle plan and eventually the battles were more strategically planned and executed.

American colonists were doomed men from the start. They just existed. They did not have private interests, any direction or ties to or property of their own. Their one purpose in life eventually became the revolution. Their sole purpose was to sever all ties with the social order, the laws and the world. Once they had done this, they were now the enemy.

The original colonies lacked any professional army. Each colony assembled its own militia. These minutemen were inadequately armed, had little or no training and lacked any type of uniforms. These units served only for a few weeks or a month at a time, were reluctant to travel far from home, were unavailable for extended operations and lacked any sort of training or discipline of experienced soldiers.

The colonists adapted to the situation and their lack of training and leadership. Their plan of attack was to hide behind rocks and trees. They laid in ambush for the British soldiers in their blue and white uniforms, who proudly marched down the roads and across open fields in a straight line formation. This type of frontier battle tactic claimed many British soldiers and resulted in a minimal amount of minutemen casualties.

The war produced very few heroes that were noted for acts of heroism from the ranks of the minutemen militia. Of course, there were those leaders that were seen as heroes from the minutemen's point of view, but were seen by the British as a person who knew nothing about leadership or what it took to perform in battle or war. The British viewed the militia as an unorganized group of farmers and storekeepers who could be defeated quickly without any problems.

Throughout history the leaders of the Revolutionary War, whether they were military or politicians, were considered the main heroes of the war because of their expertise in military or political matters. Here are a few of the many heroes that we came to know for what they did during the war. Regardless of their contributions to the war effort, some will surprise you and some are not at all as they seem.

Daniel Morgan

He was captain of the long rifles and a frontiersman who led a collection of hunters, farmers and misfits in several skirmishes and battles. He was also present at Saratoga and later led an independent campaign that culminated at the Battle of Cowpens where he destroyed the British Deviate Banastre Tarelton. Unfortunately he was part of the forces to put down the Whiskey Rebellion, even though he was barrel deep in the rum that bears his name.

Henry Lee III

He was called "Light Horse Harry" and was a flashy cavalry commander during the war. Although not known for any great specific accomplishment, he led his forces in the southern campaign. He was best known as the father of Civil War, General Robert E. Lee.

Alexander Hamilton

Hamilton served as George Washington's Chief of Staff for the last four years of the war and did a good job to keep the supply trains running to the troops. Later, famous for his political accomplishments as a founding father, it was too bad that he was the recipient of Aaron Burr's well-placed .56 caliber dueling pistol bullet. He died the next day.

William Alexander

Alexander was one of Washington's favorite generals. Washington put him in command of his entire Army while he (Washington) was away on personal business. Though he was a brilliant military man and was quite proficient in running the Army, he was a prominent drinker who loved to entertain the elite.

George Washington

First father of our country, first in war, first in the hearts of his troops, first to win independence and first to establish our system of government. Though he was first in many things, he was not first with Martha. Martha had been married before and had four children when George met her. The United States, as we know it, would not have started without him.

Were these men heroes? Of course they were. What about the ordinary citizen, the farmer and the 13 year-old boy who were given a rifle and told to stand in line on the battlefield to face the enemy and know that death was inevitable. In the annals of the Revolutionary War, few ordinary soldiers were recognized for any specific heroic feat, they just performed a duty. There were no awards or metals given for any sort of heroic action in this war. Whatever recognition given, or noted, was given to the generals and leaders who led and planned the battles. Are these battles and the men who participated in them heroes as well? Of course they are.

The Civil War

More than three million people participated in the Civil War and 1522 of them earned the Medal of Honor. This number is a huge percentage considering 3475 medals totally have been awarded for all wars since its inception. For one reason or another, some of them were awarded for acts or situations that were anything but heroic. The balance were rightfully awarded for heroic actions above and beyond the call of duty. Even though "above and beyond the call of duty" is one of the prerequisites for the Medal of Honor, it was not so in the Civil War. I am not saying that of the 1522 awarded that none were for acts of heroism, because that would not be true. What I am saying is, of the medals awarded, some were awarded for actions or reasons that were other than heroic.

The 27th Maine Volunteer Infantry Regiment

Secretary of War, Edwin M. Stanton, promised a Medal of Honor to every man of the 27th Maine Infantry Volunteer Regiment who extended his enlistment beyond the agreed upon date set forth at the time. The Battle of Gettysburg was imminent and 311 men of the regiment volunteered to serve until the battle was resolved. The rest of the men from the regiment returned to Maine. The 311 volunteers soon followed. The volunteers arrived back in Maine in time to be discharged with the men who had earlier returned. Since there seemed to be no official list of the 311 volunteers, the War Department studied the situation and forwarded 864 medals (the total amount of men in the regiment) to the commanding officer of the regiment. The commanding officer only issued medals to the volunteers who stayed behind and retained the others on the grounds that if he returned the remainder to the War department, the War Department would try to re-issue the medals.

This promise of medals to these 311 volunteers was just wrong, but at that time the prerequisites that warranted the award were few and far between. The soldiers who actually earned the award under a "above and beyond the call of duty" scenario and may have fought to near death to earn it must have felt as if the medal was anything but special. It was awarded to these men just for re-enlisting for a short period of time.

Revoked Medal of Honor Awards

In 1916 a board of five generals on the retired list, convened by law to review every Army Medal of Honor awarded. The board was to report any Medals of Honor awarded, or issued, for any cause other than distinguished service. The commission led by Nelson Miles identified 911 awards for causes other than distinguished service. This included the 864 medals awarded to the 27th Maine Volunteer Infantry Regiment, 29 who served as Abraham Lincoln's funeral guards, six civilians (including Dr. Mary Edwards Walker, the only woman to have been awarded the medal and Buffalo Bill Cody), as well as 12 others. Dr. Walker's medal was restored by President Jimmy Carter in 1977. Cody and four other civilian scouts who rendered distinguished service action and who were considered by the board to have fully earned their medals had theirs restored in 1989.

Were these who actually received this medal for actions or situations other than for heroic acts any less a hero then those who did? They all fought in the war and all faced death on an everyday basis. So who are the actual heroes? They are all heroes in their own way, no matter what they did to earn the title.

Unsung Heroes

Unsung heroes appeared throughout the war doing what came naturally. They endured hardships every single day during the war, but they persevered. The farmer gave up his land either voluntarily or it was taken by force. His food crops were taken to fuel the war. At times his personal possessions were taken or ceased. His livestock was confiscated either for work or for food to feed the troops. In some instances he may have been paid by the military for them but either way, they were part of the war effort.

He would have to join the very army that stripped him of all of his possessions in an effort to survive after all of these sacrifices. All he had left was his family and his life and his life was not assured from this point on. He was a hero whether he knew it or not.

Many unsung heroes did their part in the war effort, some by choice some not by choice. Many civilian doctors were thrown into the war because of their profession. Soldiers would be injured and doctors were needed to treat the wounded. They endured long tiring hours and most times under horrific conditions, with little or no medical supplies, to tend the wounded. Death was an every hour occurrence for these doctors and surgery without anesthesia was common because there was none.

Some of these doctors were well-educated, formidable physicians before they were sucked into the war. Now they amputated limbs that may have been saved under different situations, sometimes many per hour. In long battles the wounded came so fast that there was no time to perform long surgeries. The average length of a surgery was fifteen minutes. If it was determined that the man was wounded to the point that he may die, or lose the limb if a more in depth surgery were needed, then in most cases he would either die or the limb was amputated. The makeshift operating rooms where cut and patchwork surgery was performed were so common that a soldier would rather die of his wounds on the battlefield then go to one of them.

Nurses assisted the doctors in the operating rooms. Most of these nurses did not have formal medical training and they, like the doctors, were sucked up into the war. Those who did have any formal medical training were unable to take advantage of their training. They assisted the doctor when possible, but most times they tended to the incoming wounded to determine who was to be next on the table. The nurse and the doctor determined who would die and who may not.

Mary Edwards Walker was a doctor in civilian life and was an Acting Assistant Surgeon for the U.S. Army. Dr. Walker served in the war from 1861 through 1864 and worked in many of the key battles of the war. She devoted herself to the sick and wounded both in the field and in the hospitals, to the detriment of her own health. She was also captured and endured the hardships of a prisoner of war. Though she did not perform an actual heroic feat, she was awarded the Medal of Honor for exemplary meritorious service as she served as a doctor during the war. Dr. Walker is the only woman to be awarded the Medal of Honor. Though Dr. Walker was awarded the Medal of Honor, I place her in the unsung hero category because of her accomplishments and gender.

On November 30, 1864 the bloodiest hours of the Civil War took place in Franklin, Tennessee at the Carter farm on the edge of town. The five hour battle involved more than 53,000 troops and was one of the few night battles in the Civil War. It was also one of the smallest battlefields of the war (only two miles long and 1 ½ miles wide). The main battle began around 4:00

p.m. and ended around 9:00 p.m. As the battle raged outside of the Carter House the Carter family, along with members of the Lotz family from across the road, took refuge in their basement. Twenty-three men, women and children (many under the age of twelve) were safely protected while the horrible cries of war rang out above them. What was seen outside of the house after the battle was horrific. The front yard and the grounds around the house were covered with more than 5,000 bodies of the dead and wounded. It has been told that the ground was not visible because of the bodies. In the five hour battle 9500 troops were killed and 4800 were seriously wounded.

After the battle like so many homes in Franklin, the parlor of the Carter House along with the Lotz house were converted into a Confederate field hospital and witnessed many surgeries and amputations. Doctors worked around the clock performing surgeries on makeshift operating tables that were interior doors taken from the bedrooms. The tables (doors) were placed near the windows for two reasons; first because of the available light needed for surgery and second a place to throw the amputated limbs and body parts. The window sills were nearly five feet from the ground outside and the limbs that had been thrown out of the window were level with the sill. The Carter and the Lotz families were just some of the unsung heroes from Franklin, Tennessee. The doctors, nurses, dead and wounded made up the rest.

Surprisingly Young Unsung Heroes

In the Civil War unsung heroes came from different cultures and covered a range of ages. Ralph E. Reed, David Auld, Robert Henry Henderson, Johnny Cook, Orion Howe, Clarence McKenzie, Johnny Clem and Willie Johnson all had one thing in common; they were not soldiers, nor did they fight for either side, but they did participate in many different battles. They ranged from 9 to 18 years of age. Of these mentioned one was 18, one 16 and another only 13. The rest were under the age of 11, but all were battle veterans. These young men and boys were drummer and bugler boys. In the Civil War it is estimated that some 250,000 underage boys were involved, or enlisted, to become part of the war.

These young war participants were a vital part of the units in which they served. The drums and bugles played an important role on the battlefield communications system with various drum rolls and bugle reports.

Johnny Cook was 13 when he served as a bugler with the 4th Artillery. At 15 he was in the Battle of Antietam on the 17th of September, the first major battle of the Civil War. In the fight Johnny Cook saw the artillery gunners shot down and bravely ran to take their place. He was awarded the Medal of Honor for his assistance to help fend off three attacks. Later Johnny joined the Navy and served until the end of the war. Johnny Cook died in 1915.

Orion Howe served as a drummer with the 55th Illinois Volunteers at the Siege of Vicksburg which was considered the turning point in the War. Orion was shot in the leg and despite his wound he carried an urgent message for much needed ammunition to General Sherman's headquarters.

Robert Henry Henderson was 10 years old when he was caught up in the war as the fall of Fort Sumter took place in April 1861. Robert dreamed so much of war and to go into battle to become a hero that he lacked discipline. He ran from school and ignored his education to the point that he was unable to write his name. He enlisted in the Jackson County Rifles at 10 and the rifles were ordered to Ft. Wayne near Detroit to become C Company of the 9th Michigan Infantry in the Union Army. His unit moved from one place to another and two years later, at the age of 12, he formally enlisted in March of 1862 in Company C and was posted to Murfreesboro, Tennessee to guard the old courthouse. In July his unit was captured and he became a prisoner of the Confederacy. Robert was released because of his age and wounds that he sustained. Because of his parole captivity situation, he was disallowed the opportunity to fight the Confederacy. He then signed on again with the Union. This was not an uncommon practice for young boys who faced this situation.

An 11-year-old drummer boy of the 14th Connecticut Regiment was filling a coffee pot by a narrow stream when three Confederate soldiers came upon him. Instead of panicking, he immediately ordered them to surrender. They thought that he was not alone and they did as ordered.

Clarence McKenzie became the drummer of the Brooklyn 13th Regiment at 12. He was killed at Annapolis, Md. and was buried in Brooklyn's Greenwood Cemetery.

Willie Johnson from St. Johnsbury, Vermont was a drummer boy in Company D of the 3rd Vermont. His service in the Seven Days Battle retreat in the Peninsular Campaign was exemplary. He was the only drummer in his division to come away with his "instrument", which by no means was a trivial accomplishment. As a result he was awarded the Medal of Honor by recommendation of his division commander therefore he became the youngest recipient of the highest decoration at the age of 13.

Of all the drummer and bugler boys in the war, the most famous of all was Johnny Clem. He was not officially a member of the Army, though the officers put money toward his upkeep of $13 per month.

He was born in Newark, Ohio in 1851 and was 10 years old when the war began.

He left school classes to drill with the 3rd Ohio Volunteers where the soldiers provided him with a gun. He was refused by many of the passing regiments but managed to attach himself to the 22nd Massachusetts that adopted him as a drummer boy. He was given a shortened rifle and a specially made uniform in his size.

On September 20, 1863, at the age of 12, he was officially allowed to join the US Army and receive pay. He participated in many battles before that time. In September at the Battle of Chickamauga the Union forces were in retreat when Clem, rather than submit, shot a Confederate Colonel who had demanded his surrender. Clem was captured but managed to escape. Clem was captured once again in October by a unit of Confederate Cavalry on train duty. The Confederacy used him as propaganda to show the condition that the Union must be in, to send babies to fight. Clem was exchanged for a Confederate prisoner a little later.

He decided to change his name to John Lincoln Clem. In January of 1864 he was assigned to General Thomas' staff as a mounted orderly and stayed in the Army until discharge in September 1864.

After the war President Grant nominated Clem to become a student of the US Military Academy at West Point. However, due to his lost years at school, he repeatedly failed the entrance exams. Grant overlooked this and in 1871 appointed him second lieutenant. Clem served in the US Army until 1915 as the last Civil War Veteran that attained the rank of Brigadier General. The one-time drummer boy is buried in Arlington Cemetery.

At this time we look for a sound reason and explanation as to why children go to war; sometimes to save their families lives, for the elements of honor and glory, revenge or being forced. In simple practical terms they fight for the same reason as adults.

Women In The Civil War

In the turbulent years of the Civil War women who did not have the right to vote, own property or had few civil liberties of their own, unified in support of the war efforts and assumed an active role. The determination and devotion of these women who served their country was astounding but well overlooked as being an accomplishment.

Women may not have had a voice in the political process or a part in the military actions of the day, but the union ladies of Greenup, Kentucky did not hesitate to show their patriotism and voice their opinions.

Through a series of meetings and letters to a state representative, the women's demands to provide help in the war effort were felt and much appreciated, even in Washington, DC.

In September 1862 General George Morgan and his 10,000 man strong 7th Division marched from Cumberland Gap to the Ohio River in 16 days. They arrived starved and naked at Greenup on October 3, 1862. The soldiers soon found that the women had baked bread for them. One woman, Mrs. Ross, baked more than 200 pounds of fine flour and then, as though it was not good enough, spread it out before the 26th Brigade. She was anxious to do even more.

Women found ways to support the soldiers during the organization and formation of the 14th Kentucky Infantry in the fall of 1861. Charlotte Culver, a well-to-do widow from Catlettsburg, made her house and property available to the regiment as a headquarters and hospital.

October 1861 Elizabeth Pennington, wife a local prominent miller, baked bread for the regiment for two weeks because the soldiers did not have a way to bake bread.

Soon the women cared for the sick. An act passed by Congress on August 3, 1861 made their participation, although limited, possible.

Many women also organized Soldiers Aid Societies, as in Greenup, KY.

These organizations were instrumental in gathering and distributing items to the hospitals and battlefield. Not only would the women knit socks and mittens, make uniforms, distribute blankets and reading material, but also took it upon themselves to raise money to support their organizations. Through the efforts of these women and their organizations hospitals were supplied with necessities such as crutches, bandages and linens, as well as clothing, pillows, bedding and even furniture. They also provided fresh fruit and vegetables, eggs, condensed chicken, milk and pickles to aid the recovery of the sick.

Circumstances of the Civil War forced women to abandon their more traditional roles as wives and mothers who tended to their families. Now women worked to manufacture arms, ammunition, uniforms and other supplies for the soldiers. On the home front they took the place of their husbands. They tended to their farms, plowed, planted and harvested crops and took care of the livestock. Since often they were left on their own without the protection of their husbands and sons, they were subjected to raids by contending armies, as well as guerrilla bands.

The Civil War affected all women in this country on some level and even more so in towns such Greenup, KY. Regardless of what roles women assumed, their contributions made an enormous impact and proved invaluable to the war effort. Their patriotism and sacrifices, as well as their triumphs, should never be forgotten. They are truly unsung heroes.

Blacks In The Civil War

Many blacks served in different capacities in the Civil War. Twenty-five black men earned the Medal of Honor in the Civil War. As great as this feat was, the following people performed feats just as important.

The protection of vital military information and secrets was of the utmost importance in the Civil War. The military had enough problems to make sure that their battle plans were performed with accuracy and with the utmost of secrecy. A leak of vital battle plans would result in a disaster on the battlefield. So extreme measures were taken to protect this material. As well as the Union and Confederate Intelligence Services did to protect vital information, they did not figure on John Scobell as the source of stolen plans and secrets.

John Scobell was considered just another Mississippi slave in the Confederate circles he navigated; he sang, shuffled, was illiterate and completely ignorant of the Civil War that took place around him.

Confederate officers thought nothing of leaving important documents where Scobell could see them or discuss troop movement in front of him. Whom would he tell? Scobell was only the butler, a servant, the deckhand on a rebel sympathizer's steamboat, or the field hand that would belt out Negro spirituals. In reality, Scobell was not a slave at all.

He was a spy sent by the Union Army. He was one of a few black operatives who quietly gathered information in a high-stakes game of cat-and-mouse with the Confederate spy-catchers and slave masters who would kill them on the spot. These unsung Civil War heroes were often successful to the embarrassment of the Confederate leaders, who never thought their disregard for blacks who lived among them would become a major tactical weakness.

"The chief source of information to the enemy," General Robert E. Lee, commander of the Confederate Army, said in May 1863, "is through our Negroes".

Little is known about the black men and women who served as Union intelligence offices other than the fact that some were former slaves and servants who escaped their masters. Others were Northerners who volunteered to pose as slaves to spy on the Confederacy. There are few references to their contributions in historical records, mainly because Union spymasters destroyed all documents and records to shield them from Confederate soldiers and sympathizers during the war and not to mention the vengeful whites afterward.

Harriet Tubman is the most recognizable of these spies. She would sneak down South repeatedly to gather intelligence for the Union Army while she also led runaway slaves to freedom through the Underground Railroad. Often disguised as a field hand or poor farm wife, she led several spy missions into South Carolina while she directed others from Union lines.

Mary Elizabeth Bowser was born a slave to the Van Lew family that freed her and sent her to school. Bowser excelled in school and upon completion she returned to Richmond, where Mary ran one of the war's most sophisticated and complex spy rings.

Somehow Van Lew got Bowser a job inside the Confederate White House as a housekeeper. Bowser then proceeded to sneak classified information out under Confederate President Jefferson Davis' nose.

The jobs that were performed by these spies gained them little rewards other than a better life style than the average slave. They were under constant pressure from both armies to keep the most current and important information flowing. If discovered, it meant certain death. Definitely heroes in the eyes of all.

Photographers of The Civil War

There were a large number of battles and other scenes of the American Civil War and collectively the photographers have provided the world with a photographic firsthand account of this period in American history.

There were many photographers from both the Union and Confederate Armies, who chronologically cataloged the Civil War in photos. These men are just a few of the heroes that went into battle without a weapon.

Mathew B. Brady

Mathew B. Brady, because of his mastery of his job, was the most prominent photographer of the Civil War. Brady would later spend his entire accumulated savings from operating a photography studio in New York City to take pictures of the war.

At the start of the war in 1861, Brady organized his employees into groups in order to spread them across the country and to get work. Brady at his own expense provided carriages to all of the parties which acted as rolling darkrooms to develop the photographic plates into pictures. The total cost was about $100,000.

The First Battle of Bull Run provided the initial opportunity to photograph an engagement between opposing armies. Brady was almost killed at Bull Run and in the confusion of battle got lost for three days. He eventually made his way back to Washington and was nearly dead from starvation.

After the war Brady went bankrupt and was forced to live off his friend's generosity. The government bought his collection of 5700 plates for a paltry $25,000, much lower than the $125,000 he asked. He once said that long after he was dead, his work would be appreciated. Some familiar with Brady's work felt that he was just as much of a hero as those soldiers who went to war. He died in poverty and total isolation in 1896.

Alexander Gardner

Alexander Gardner was born in Scotland in 1821 and excelled at learning photography. Brady heard of Gardner's achievements and invited, and paid, Gardner to come to New York to work for him.

When the war began Gardner was appointed to the staff of General George McClellan, the commander of the Army of the Potomac. He was given the honorary rank of captain and in this capacity photographed the battles of Antietam, Fredericksburg, Gettysburg and the Siege of Petersburg.

Gardner also photographed Mary Surratt, Lewis Powell, George Atzerodt, David Herold, Michael O'Laughlin, Edman Spangler and Samual Arnold who were arrested for conspiring to assassinate Abraham Lincoln. He also took photographs of the executions of Surratt, Powell, Atzerodt and Herold who were hanged at Washington Penitentiary on July 7, 1865. He photographed the execution of Henry Wirz, Commanding Officer at the infamous Andersonville Prisoner of War camp in Georgia four months later.

In 1865 Gardner was charged with photographing Lincoln's assassins. He published his classic two-volume work, Gardner's Photographic Sketch Book of the Civil War, in 1866. Each book contained 50 hand-mounted original prints. However, it was not a great sales success.

James F. Gibson

James F. Gibson was probably the least known of the Civil War photographers. He was born in New York City and eventually learned the art of photography from Brady. Gibson eventually photographed General McClellan's Peninsula Campaign, Seven Days Battles, Battle of Gaines Mill and Battle of Malvern Hill.

Southern Photographers

Many photographs were taken by Southerners, but most were lost to history. The natural disappointment in the South at the end of the war was such that photographers were forced to destroy all negatives and available original photographs that could be found. Fortunately, thousands escaped the destruction.

George S. Cook

George S. Cook was one of the foremost Confederate photographers, thanks to his recording in photos of the gradual destruction of Charleston and Fort Sumter by enemy action. He even photographed the naval action of ironclads at Fort Sumter. Unfortunately most of Cook's photographs were lost in a fire in 1864. In 1880 Cook bought the businesses (and the negatives) of the photographers who were retired and moved from the city. Thus he amassed the most complete photographic collection held in one location of a former Confederate capitol.

Robert M. Smith

Confederate Lieutenant Robert M. Smith was captured and imprisoned at Johnson's Island, Ohio. He is unique in that he was able to secretly construct a wet-plate camera using a pine box, pocketknife, tin can and a spyglass lens. Smith acquired chemicals from the prison hospital to use for photographic processing. He clandestinely used the camera to photograph other prisoners at the gable end of the attic of the cellblock four.

Photographers would often follow the armies from one location to another and inevitably into battle to get the photographs.

The results of the efforts of all Civil War photographers can be seen in almost all history texts of the conflict. In terms of photography, the American Civil War is the best-covered conflict of the 19th century. It set the standard of photojournalism of World War II, the Korean War and the Vietnam War.

These men can be idolized as combat heroes, without weapons, who put their lives on the line to secure a part of history.

Surprising Unsung Heroes

Performance of an act of heroism usually will result in becoming a hero in most cases. This may be one of those situations where just being there will warrant being a hero.

In the Civil War one million five hundred thousand (1,500,000) horses and mules served in many different capacities. Most of them were procured and then trained for field use. Many were used as personal mounts for officers in all cavalry and infantry units. Both horses and mules performed some of the same tasks. They pulled supply wagons that delivered supplies to units in the field as well as encampments. They also pulled the artillery pieces and ammunition wagons to and from the battlefields. They were used to drag lumber from the forest that was needed to build bridges across rivers and streams and buildings at their headquarters and main encampments.

These animals were treated anything but humane. They were an expendable commodity. If one died, another one would take its place. Thousands died of mistreatment and malnutrition. Some were consumed by the troops when food was in short supply. The troops showed no favorites when it came to food, especially when there was none.

Of course the majority of them were killed on the battlefields. When a soldier was knocked from his mount and the horse was unhurt, the troops used them as shields to advance. When the enemy saw this tactic it meant that the animal was a target.

Are the horses and mules of the Civil War heroes? Why are they not considered heroes? They carried their rider into battle and died on the battlefield just as the soldier did? These animals were definitely heroes.

Inanimate Heroes

Can an inanimate object be, or become, a hero? This has been argued for many years and still there is no real answer. The decision on whether it is, or is not, or possibly could be, is determined by the individual. This individual says, "Yes".

Tinclads were the heroes of the Union Brown Water Navy. These modest, but effective, gunships that were developed by the Union Navy tipped the balance of power on the rivers in the Civil War. They were faster, more maneuverable and more versatile than the Ironclad gunboats and the Confederacy had no response to them. The union used them to patrol western rivers, as transports, ground support for the artillery and anything else needed.

In a time when the Union had to move a large number of troops and supplies up the rivers quickly, these Tinclads could not have come at a better time in the war.

James B. Eads

After the outbreak of the American Civil War in 1861, Eads was called to Washington to consult on the defense of the Mississippi River. Soon afterward he was contracted to construct the City Class Ironclads for the United States Navy and produce seven such ships within five months; The

St Louis, The Cairo, The Carondelet, The Cincinnati, The Louisville, The Mound City and The Pittsburgh. The river Ironclads were a vital element in the February-June highly successful Federal offensive into Tennessee, Kentucky and Upper Mississippi. Eads corresponded frequently with a Navy officer of the Western Flotilla and used their "combat lessons learned" to improve vessels during post-combat repairs and build improvements into succeeding generations of gunboats. He would build more than 30 river Ironclads by the end of the war. All senior officers in the Western theater, including Grant and Sherman, agreed that Eads and his vessels had been vital to early victory in the West.

The first Tinclad was a converted riverboat. It was purchased by the Union for $24,000 and the conversion was finished in early 1862. It served mainly as a patrol boat in the beginning and was used extensively as a way to transport troops and supplies quickly to locations that would take much longer if traveled by land. Although this was a great asset to the Union, the Tinclad had its disadvantages. The boat had a wooden hull and the draft of the boat was four and one half feet. The craft had little trouble to negotiate the rivers under normal river conditions. The Union found that the boats needed more armor. Anything above the waterline was covered in steel plate armor which caused a problem when the river conditions were anything other than normal. In the summer months the river levels would drop and in some cases significantly. This limited where the boat could and could not go. Sandbars were everywhere when the rivers were low and submerged tree trunks were an extreme danger to the wooden hull. The addition of more protective steel plating caused the boat to sit lower in the water, which limited even more what the boat was able to do.

Add six 24 to 30 pound cannons to the weight of the boat and in low water conditions it floundered. This craft carried one captain and 16 crewmembers. Living conditions on the boat were quite good because nearly all of the boat sat above the water line.

The Ironclad soon followed. The problem with the Iron Clad was that most of the hull was submerged below the water. The only portion of the boat that was above water level was a few inches of its massive deck and the gun turret that encased the boats armament. The Ironclads were seagoing vessels and only went into the mouths of the rivers that flowed into the Atlantic. This ship was a formidable adversary against the Confederate ships that patrolled the coastline because they were heavily armed and more than sufficiently covered in armor plating. It did not fare well in rough water because initially it was constructed for seagoing use.

The ship was propelled by two massive coal-fired steam boilers. Living quarters were minimal because space onboard was needed to store the coal that was needed to move the craft and ammunition to protect it in battle.

Meals on the boat consisted mainly of soup and bread and bad soup at that. The crew of 59 men and officers were all below while underway. Because of the constant movement on patrol, there were very few instances when the crew was allowed on deck. The temperature inside was near 130 degrees and what fresh air that was inside was quickly consumed by the boilers. Living conditions onboard were atrocious and most times the crew's attitude reflected this.

As bad as these issues were for the crew, these ships made the turning point for the Union Navy in the war. Then came the Merrimack.

Even though these ships were inanimate object's they and their crews were heroes of the Civil War.

World War I

More than sixteen countries from across the globe were in WW I. Seventy million military personnel, that included sixty million Europeans were mobilized in one of the largest wars in history. More than nine million combatants were killed, largely because of the great technological advances in firepower, without corresponding advances in mobility. It was the sixth deadliest conflict in world history.

World War I produced thousands upon thousands of heroes recognized for heroic feats in combat. Alvin York was one of the most noted heroes of the war. On June 17, 1917 at the age of 29 York registered for the draft, as all men between the ages of 21 and 31 did. When he registered for the draft he answered the question, "Do you claim exemption from the draft (specify grounds)?" "Yes. Don't want to fight." He appealed when his initial claim for conscientious objector status was denied. In World War I conscientious objector status did not exempt one from military duty. Regardless of his strong religious beliefs, he was drafted into the Army. Even though York knew and felt that killing another man was against all of his beliefs, he felt compelled to go to war like all others that he knew and respected. He talked with his company commander about how he felt and was granted a 10-day leave to visit home. He returned convinced that God meant for him to fight and would keep him safe. He was as committed to his new mission as he was to pacifism.

On October 8, 1918 York's actions earned him the Medal of Honor during an attack by his battalion to secure German positions along the Decauville rail line in northern France.

At this point in this story I will not go into detail of York's accomplishments which earned him the Medal of Honor, but only give the basics of the action.

York led seven men who charged a German machine gun position that had his platoon pinned down and had caused many casualties. York was separated from the very men he led, but continued to attack and silence one machine gun position after another. Six German soldiers in a trench near York charged him with fixed bayonets in the assault. York had fired all the rounds in his M1917 Enfield Rifle, but drew his .45 Colt Automatic pistol and shot all six soldiers before they could reach him. York and his seven men captured 132 enemy soldiers (including three officers), killed 28 German soldiers and took 32 machine guns by the end of the engagement. For his actions York received nearly 50 decorations.

I thought that it would be appropriate to mention York's heroic actions in the war. Though he was a hero, the following were just as much a hero in their own right as was York.

The WW I Heroes of Technology

The evolution of true technology may have started in the Revolutionary War and progressed quite quickly through the Civil War. The standards of weaponry for the early wars was sufficient enough for the type of warfare waged at that time but with a more advanced type of war, it was inevitable that technology would have to take a large step forward.

There were many innovations that came to be in World War I. Here are just a few examples of improvements and inventions that came about in this war.

Hugo Junker

Hugo Junker experimented with aircraft and how to improve the aircraft body to improve flight and maybe give the pilot more protection. He experimented with different materials to find one that was suitable enough to stand up to the harsh conditions that the aircraft experienced in flight, landings and possible gunfire.

In 1915 Junker designed the first all-metal plane. He finally settled on a form of lightweight sheet metal to replace the heavy canvas for the covering on the fuselage. He also replaced the wooden framework on the interior of the plane body, which eliminated much of the steel cables that basically held the plane's frame together. The metal stood up to all conditions, even though it did not provide much protection against small arms and machine gun fire that the aircraft were subjected to on every combat flight. Though the metal added additional weight to the aircraft, it required very little maintenance like the heavy canvas and wood framing that was susceptible to gunfire and an occasional fire. His ideas were too far advanced for his time and the production of the plane did not start until 1918.

Roland Garros

Roland Garros was a pilot in the war and found out quickly that a machine gun mounted on the front of his aircraft behind the propeller damaged the propeller when fired. As the propeller turned and the bullets passed through the propeller blades, the bullet would strike the propeller blades which sliced off pieces of the blades. The damage seemed to be only on the edge of the propeller blades. Garros determined that if he added a piece of steel to the edge of the blades it may deflect the bullet enough to save further damage to the propeller and have the bullet still do its job to strike the enemy aircraft.

On Garros' first flight he encountered an enemy aircraft. He headed straight for the German plane, something that was not done in air combat because of the bullet propeller issue. Usually when two opposing aircraft met each other there was a dogfight that involved one or both trying to maneuver into position where they could fire at each other by using pistols or another small firearm.

That day Garros shot down his first enemy plane and the next day he shot down three more. His idea of a deflector plate actually worked.

Unfortunately soon after these victories, his plane's fuel line was shot by ground fire and he had to make an emergency landing in enemy territory.Before he could set his plane on fire to keep it from falling into enemy hands, he was captured. The Germans recovered the aircraft, discovered the deflector plates, and soon used them on their own aircraft.

Later on in the war a cam-gear was devised that tied the rotation of the propeller with the sustained fire from the machine gun and therefore allowed the bullet to pass through the propeller without striking the blades. A formidable accomplishment to say the least.

Other timely innovations that were introduced during the war were:

Aircraft Carriers
U-Boats
Tanks
Howitzer Artillery Pieces
Parachutes
Trench Railways
Aerial Photography
Gas Masks
Field Telephones
Wireless Communication
Hydrophones
Steel Helmets
Bolt Action Rifles
Light Automatic Weapons
Submachine Guns

The Browning Automatic Rifle
Flame Throwers
Smokeless Gun Powder

There were many, many more inventions that surfaced in the war but none as important as the aforementioned. The innovators of these inventions, even though not involved in combat or even a member of the military, are certainly heroes in their own right.

The Land-Ships of World War I

There were many ideas to develop a vehicle that could travel over land and deliver substantial firepower while protecting the ground troops at the same time.

The Land-Ship Committee and the newly formed Inventions Committee agreed to the specifications for a new machine. This included: (1) A top speed of four miles per hour (2) The capability of a sharp turn at top speed (3) A reversing capability (4) The ability to climb a five-foot embankment (5) The ability to cross an eight-foot gap (6) A vehicle that can house ten men, two machine guns, a two-pound gun plus provide enough room for ammunition.

The first tank prototype was called Little Willie. It's Daimler engine was grossly under powered, had 12 foot long tracks, weighed 14 tons and could only carry a crew of three instead of the agreed upon 10 men. The speed dropped to less than two miles per hour over rough ground and most importantly of all, it was unable to cross broad trenches.

Although the performance was disappointing, the committee was convinced that, when modified, the tank would enable the allies to defeat the Central Powers.

The first large engagement of tank warfare involved 12 Divisions of personnel and 49 tanks in 1916. There was so much secrecy in the development and manufacturing of the tank that the Germans were completely taken by surprise, so much so, that two miles of German-held territory was captured. However a large portion of the tanks broke down and the Army was unable to hold onto its gains.

Within the next year improvements were made to the tank which made it a formidable piece of military might. Unfortunately the Germans developed their own version of the tank. Tank warfare was born.

The Unsung Heroes of World War I

There were millions upon millions of unsung heroes in World War I. Many were just individuals who tried to survive from day to day. Many units and/or specific groups of people who specialized in one certain field participated in the war.

Harry L. Wingate

Harry Wingate was a pilot and flight instructor from 1917 to 1919 in France. He did not see combat as a pilot because the Armistice of WWI was signed just as he was to be assigned to an aero squadron at the front. This in no way diminishes his contribution to the war effort according to military sources. He was credited with training 75 percent of the pilots who fought in WWI, an amazing accomplishment.

Black Soldiers In World War I

Thousands of African Americans participated in World War I. Because of their ethnicity they were placed in positions or units that failed to recognize them as true fighting heroes, either for a single heroic action or collectively as a group or unit.

At first African American units were to be used as part of the French divisions. The Harlem Hell Fighters (so named) fought as part of the French 16th Division and earned a unit Croix de Guerre for bravery in action against the German forces at Belleau Wood, France.

The Red Hand Division

The 371st Infantry Regiment was an all black army unit that fought during WWI. This African American unit was trained at Camp Jackson in Columbia, S.C. and fought alongside French troops and used French weapons and equipment. The 157th French Division, to which they were attached, became known as the "Red Hand Division" under the command of French General Goybet. Enlisted members of the 371st earned 12 American Distinguished Service Crosses and 89 French Croix de Guerres. The 371st fought consistently on the front lines and consistently drove back the German lines that resulted in the unit's flag "Red Hand" insignia. Though the Red Hand Division's accomplishments and heroic feats were many, notoriety for such bravery went virtually unnoticed in America.

Henry Lincoln Johnson

Henry Johnson enlisted in the Army on June 5, 1917 and joined the all black New York National Guard Unit, the 15th New York Infantry. When the unit was mustered into federal service it was renamed the 369th Infantry Regiment based in Harlem, New York and was assigned to the French Command in WWI. Johnson arrived in France on January 1, 1918.

On May 14, 1918 Private Johnson came under attack by a German raider party while on guard duty. Johnson displayed uncommon heroism when he used his rifle and bolo knife to repel the Germans and thereby saved a comrade from capture, plus he saved the lives of his fellow soldiers. The act of valor earned him the nickname "Black Death" as a sign of respect for his prowess in combat.

Johnson was the first American soldier in WWI to receive the Croix de Guerre with Star and Gold Palm from the French government and also received the Distinguished Service Cross and Purple Heart from the U.S.

Johnson died in New Lenox, Illinois at a veteran's hospital in 1929. He was penniless, estranged from his wife and family and without official recognition from the US government. He is buried in Arlington National Cemetery. June 1996 Johnson was posthumously awarded the Purple Heart by President Bill Clinton. February 2003 the Distinguished Service Cross (the Army's second highest award) was presented to Herman A. Johnson, one of WWII Tuskegee Airmen, on behalf of his father.

Eugene Bullard

A largely unheralded hero of the Lafayette Flying Corps was Eugene Jacques Bullard. He was an African American from Columbus, Georgia who would become the first African American pilot. The son of a freed slave, he left Columbus by himself to move to Atlanta while still in his teenage years. He had been told that in order to escape racial prejudice he should head to Europe, especially France. Eugene's father had pointed out to him that Bullard was a French name and that at least one of their ancestors was French. Soon he moved from Atlanta to London and then on to Paris. France had been good to Bullard and when WWI broke out, he signed up for the French Foreign Legion. He was sent to the French Army's 170th Infantry Regiment whose nickname was the "Swallows of Death". He was wounded twice at Verdun and then sent to a Parisian hospital where he spent the next six months in recuperation. His valor was recognized with a chest full of French military decorations that included the French Croix de Guerre.

On October 5, 1916 he arrived at the French Aerial Gunnery School at Cazaux on the Atlantic. He eventually obtained his pilot's license through some very difficult and long training. It was this accomplishment that would launch Eugene Bullard into history as the first ever African American aviator.

Though obtaining his pilot's license was an accomplishment, especially for a black man, it affronted no special treatment or favors when the time came to join the elite pilots group Lafayette Escadrille as one of their pilots. The day he was officially rejected was August 23, 1917.

He was eventually attached to another lesser-known unit and flew air patrols on a daily basis to the front lines. From August through November of 1917 he flew numerous patrols. He flew a few different aircraft in these patrols and on his last plane there was an insignia on it that said, "All Blood Runs Red" and his nickname became the "Black Swallow of Death". By some account, Bullard was said to have shot down two enemy planes, but it was stated that neither one was confirmed. Later he was bumped out of the French Air Force and then transferred back to the 170th Infantry Regiment of the French Army.

After the war Bullard bought a bar in Paris. In the late 1930's, prior to the outbreak of WWII, he was recruited by French Intelligence to spy on the Germans who came into his bar.

When WWII broke out in 1939, Bullard still lived in Paris and was running his bar. He became very devoted to France and tried to join the army but was considered too old. In 1940 he managed to find a way out of German occupied France and returned back to the United States.

In 1954, along with two other French veterans, he was invited by French President Charles de Gaulle to light the flame of the Unknown Soldier at the Arc of Triumph in Paris.

He died in 1961 at the age of 66 with his achievements all but forgotten.

While Eugene Bullard, as an African American aviator, is not as famous as the Tuskegee Airmen or Benjamin O. Davis Jr., he was before of all of them. Some heralded him "as probably the most unsung, unheralded hero in the history of U.S. wartime aviation". Others noted that his single-handed accomplishment was the equivalent of what all the Tuskegee Airmen combined had accomplished in World War II.

The Code Talkers

Over the past few years a group of Native Americans has been recognized for their part in World War II. Native American Code Talkers of the Choctaw Indian Tribe were involved in war well before the Navajo Code Talkers of World War II.

In 1917 Choctaw Indians were not citizens of the United States. The language the Choctaws spoke was considered obsolete. That same language later helped bring a successful end to the First World War. A number of Choctaw soldiers baffled and confused German eavesdroppers almost to the point of total embarrassment.

In war communication is an important weapon, a weapon that can be used to defeat your enemy or destroy you. In WWI the Germans were able to decipher all of the communications of the Allied Forces. Then something miraculous happened, a group of 19 young Choctaw men appeared on the scene and used their own native language to transmit messages that the German were never able to understand, let alone decipher.

Native Americans, including Choctaws, were not allowed to vote in 1924. Although years before this, they volunteered to fight for what they considered their country, land and people. According to tribal documents and military records, there were 19 Choctaw Code Talkers: Tobias Frazer, Victor Brown, Joseph Oklahombi, Otis Leader, Ben Hampton, Albert Billy, Walter Veach, Ben Carterby, James Edwards, Solomon Louis, Peter Maytubby, Mitchell Bobb, Calvin Wilson, Jeff Nelson, Joseph Davenport, George Davenport, Noel Johnson, Schlicht Billy and Robert Taylor. The men listed here were part of the 36th Division. Originally only eight men were recognized as Choctaw Code Talkers, but as the success of the use of their Native language as a "code" was recognized, others were quickly introduced into service.

Toward the end of the war the Germans tapped radio and telephone communications. Messengers were sent out from one company to another. These messengers had been dubbed runners. One out of four runners was captured by German troops. The Germans had decoded all transmitted messages up to this point in the war.

It was well understood that the Germans were masters of "listening in" to every message transmitted. While the Code Talkers were comparatively inactive, it was remembered that the regiment possessed a company of Indians. They spoke twenty-six different languages or dialects, only four or five of which were ever written. There was hardly a chance of one in a million that the Germans would be able to translate these dialects and the plan to have these Indians transmit telephone messages was adopted.

The regiment was fortunate to have two Indian officers who spoke several of the dialects and Indians from the Choctaw Tribe were chosen to perform these duties.

When the Choctaw tongue was spoken over the field telephones, the Germans stopped attacking the supply dumps and counter attacking the American troops. This is because they had no idea what the Choctaws were saying and could not effectively spy on the message transmissions. A captured German officer confessed that his intelligence personnel were completely confused by the Indian language and gained no further benefit whatsoever from their wiretaps.

The Germans had little to research in regard to the Choctaw or any Native American language. Since the Germans had been successful to decipher American-coded messages, they had some idea of how the Allies might code their secret communications. Most Americans are of the European origin. Choctaws on the other hand, are not, so the Germans had no reference to translate the native languages, especially the Choctaw dialect.

Within 24 hours after the Choctaw language was initiated, the tide of the battle had turned and in less than 72 hours the Allies were on full attack.

While more than one Choctaw soldier claimed to have come up with the idea to use the Choctaw language to confuse their enemies, the Army argued that the initiative was theirs.

It was not until many years after WWII that the Navajo Code Talkers were recognized for their accomplishments. This brought recognition to the Choctaw for what they did in WWI.

At this point it would be difficult to determine how many battles were won and lives were saved because of the Choctaw Code Talkers. They are true heroes.

The Merchant Marine Service

The United States Merchant Marines refers to the fleet of US civilian-owned merchant vessels operated by either the government or the private sector, that engage in commerce or transportation of goods and services. The Merchant Marines is responsible for transportation of cargo and passengers in peacetime. In time of war, the Merchant Marines are an auxiliary to the Navy and can be called upon to deliver troops and supplies for the military.

In wartime The Merchant Navy, as a service, has no real way to display its capabilities. There is no compulsory wearing of uniforms, any street parades, or ceremonies of any sort to attract the public or media. The only attention given to the service is when some catastrophe occurs which raises the attention of the conservationists. This always seems to be to the detriment of the Merchant Navy.

If a warship is lost by conflict or by dereliction of duty, we never hear the last of it. If a merchant ship is lost, you seldom hear of it. When a merchant ship is lost in a conflict, most times it was with all hands and left no trace.

Merchant ships of all sizes quietly come and go. They visit ports, both large and small, across the world and carry the raw materials of trade, oil and petroleum products plus the manufactured goods of industry. It is in times of conflict that the Merchant Navy finds itself as an indispensable force within the framework of military operations and even then, for safety and security reasons, a low profile is maintained.

In wartime nations with extensive global interests involved

in such confrontations look towards their merchant ships for sea lift capabilities in the transportation of their military personnel, equipment and supplies to wherever-whenever they are required. Also used to sustain the nations for the duration of the war with the necessary arms and ammunition, fuel and food and all the goods of war and then bring everyone safely back home again.

It was an accepted fact of life that the Merchant Navy would go anywhere it was sent without argument. In the Boer War it was the Merchant Navy that took all Australia's troops to South Africa and brought them home again safely. But there were no U-Boats, enemy aircraft or mines to worry about.

However in WWI there were U-Boats, mines, enemy aircraft and a lot more. In WWI the first person killed was a merchant marine when his ship, a brigantine, was sunk by gunfire from a German U-Boat in the North Atlantic.

Records show that in WWI 197 Merchant Marine ships were destroyed or sunk from different causes. Of all the causes 148 ships were sunk by German U-Boats with a loss of 629 persons. There may have been many more ships and persons lost that were not recorded. The ships of the Merchant Navy and the Merchant Marines who manned them were the unsung heroes of the seas in both world wars.

Women and The War

World War I is remembered as a soldier's conflict for the six million men who were mobilized and for the high military casualties compared to civilian deaths.

However it was also a total war where the entire nation's population was involved. Everyone involved contributed to the war effort from civilians who worked in the factories to make uniforms, guns, tanks, planes, ammunition and ships, to families with men at the front. Certainly the most prevalent group that contributed a major role in the World War I were women.

Women took on the responsibilities, not only at home,

but to replace men in offices and factories and serve in the Armed Forces. More than 25,000 women served in Europe in WWI. They helped nurse the wounded and provide food and other supplies to the military. They were telephone operators, entertained the troops and adhered to the expectations that were expected of them from society.

When most men were across the ocean to fight a war, the home front soon realized the extreme shortage of workers. Before the war women and families depended on the men for financial support. However with so many gone to war, women had to go to work to support themselves. To show support for the war, one by one, women stepped up to do a man's work for very little pay, respect or recognition for their efforts. Severe labor shortages provided the opportunity of a variety of jobs that were unavailable to women before the war. They became streetcar conductors, railroad workers and shipbuilders.

Some women took over the very source that fueled the nation, the farms. They monitored the crops, harvested them and they tended to the livestock. Women, who had young children with nobody to help them, did what they could to help also. They made things for the soldiers overseas, such as flannel shirts, socks and scarves, which was their way to show the soldiers that they too were helping and thinking of them.

While the women of the United States did their part in the war effort, countries in Europe did their part as well. In the early years of the war Britain quickly experienced a crisis with ammunition supplies. The army fired off shells faster than industry could produce them. Upon notice of the dilemma they quickly took advantage to place the women of Britain into the workforce, especially as labor in the ammunition factories. Initially a man's job, but before long, 90 percent of the workers who made ammunition were women.

The courage of the women in the munitions factories was never sufficiently recognized. They had to work under conditions of real danger to life and limb. What some of them who worked in the shell filling factories really dreaded, even more, was disfigurement caused by toxic jaundice that resulted from TNT poisoning. This turned their skin a bright yellowish color. They were nicknamed "canaries." They were quite proud of this designation as they had earned it in the path of duty.

The factories were very dangerous and unhealthy and the women only received one-half the wages of men. The women were not unionized because the Labor Union claimed that they had to hire many women to replace one man and that the skilled tasks were broken into several less skilled tasks, which therefore required a lower wage. The women felt that they qualified for the tasks at hand but they felt that even though they filled the need in manufacturing, they did not get the true respect and credit they deserved for what they did.

Eventually the women started their own union, The National Women's Trade Union League. But wages were not raised because the wages were controlled by an authority much higher than the women's union. The women had a hard time to adjust to many changes, but they persevered.

Women were viewed as important and respected for their part to fight the war. Images of women on posters and postcards were provided to the troops in battle for inspiration for what they were doing back home. The belief was when a man saw the image of a woman, he would be reminded of what he was fighting to protect. It would also give him as sense of comfort about his loved one back home.

Because of the service the women provided in World War I, it resulted in a huge push to the passage of the 19th Amendment. President Woodrow Wilson urged the Senate to reflect on the bravery of the women and their proven abilities in the offices, factories and on the front during the war. Passage of the 19th Amendment was essentially an important thing to do.

Nursing On The Front Lines

With so many men fighting in the trenches there were hundreds of wounded every day. Nurses were brought to the front lines to help with the treatment of those wounded. Because they worked on the front lines, these women ran the risk to be hit by a stray bullet or even shelled in an enemy bombardment. Many of these women were killed as they carried out their duties.

Girls as young as 16 worked as nurses and Help Wanted Ads for nursing increased by the day. Many young women volunteered to join the Voluntary Aid Detachment and First Nursing Corps. They had very basic nursing skills but they could still help the wounded soldiers in the war zone by giving them basic medical treatment. As good as their intentions were, especially with their limited medical training, some knew little or nothing about nursing and had to learn on their patients, which was a painful process for all concerned. As volunteers they were not paid.

The very first nursing groups were in charge to drive ambulances, run soup kitchens for the soldiers and tend to their needs when the soldiers had time off at the front line. Educated and trained Physical and Occupational Therapists were called Reconstruction Aides and saw service in the Armed Forces by serving in hospitals in the United States and overseas. At least three Army nurses were awarded the Army's second highest military honor, The Distinguished Service Cross.

Women not only enlisted in the Army as nurses, but many were sworn into the U.S. Army Signals Corps as operators. In late 1917 General Pershing put out an emergency appeal in many newspapers for bilingual telephone switchboard operators. Pershing wanted women to be sworn into the Army as an emergency need because, as he stated, women have the patience and perseverance to do long detailed work. He found that men in the Signal Corps had a difficult time to operate switchboards. He thought that men could serve better on the front lines and string communication lines from the trenches to the General Command. The women operators became known as the "Hello Girls."

Keeping The Home Fires Burning

Even with their new found duties women still found the time to write to their sons, boyfriends, husbands, brothers and friends who were fighting on the front lines. They sent them mementos from home such as pressed flowers from the garden, photographs and embroidered handkerchiefs. These letters proved to be essential to boost the morale of the homesick and frightened soldiers.

The suffering and deaths of many American soldiers in World War I will always be remembered and honored for their dedication and sacrifice they made for their country. The unsung heroes of WWI, the women, not only made sacrifices but also kept the industrial wheels turning and the home fires burning.

Animals In WWI

Cher Ami (French for "dear friend") was a Black Check Cock homing pigeon and was probably one of the most unrecognizable unsung heroes of the war. He was donated by the pigeon fanciers of Britain for use by the U.S. Army Signal Corps in France during World War I and had been trained by American pigeoneers. It helped save the Lost Battalion of the 77th Division in the Battle of the Argonne in October 1918.

On October 3, 1918 Charles Whittlesey and more than 500 men were trapped in a small depression on the side of the hill behind enemy lines without food or ammunition. They also began to receive friendly fire from allied troops that did not know their location. Surrounded by Germans, many were killed or wounded in the first day. By the second day just more than 200 men were still alive. Whittlesey dispatched messages by pigeon. The pigeon that carried the first message "Many wounded. We cannot evacuate." was shot down. A second bird was sent with the message, "Men are suffering. Can support be sent?" That pigeon also was shot down. Only one homing pigeon was left; 'Cher Ami'. He was dispatched with a note in a canister on his left leg, "We are along the road parallel to 276.4. Our own artillery is dropping a barrage directly on us. For heaven's sake, stop it." As Cher Ami tried to fly back home, the Germans saw him rise out of the brush and opened fire. For several moments Cher Ami flew with bullets zipping through the air all around him. Cher Ami was eventually shot down but miraculously managed to take flight once again. He arrived back at his loft at division headquarters 25 miles to the rear in just 25 minutes and helped to save the lives of the 194 survivors. In this last mission Cher Ami delivered the message despite having been shot through the breast, blinded in one eye, covered in blood and with a leg hanging only by a tendon.

Cher Ami became the hero of the 77th Infantry Division. Army medics worked long and hard to save his life. They were unable to save his leg, so they carved a small wooden one for him. When he recovered enough to travel, the little one-legged hero was put on a boat to the United States, with General John J. Pershing personally seeing Cher Ami off as he departed France.

Upon return to America, Cher Ami became the mascot of the Department of Service. The pigeon was awarded the French Croix de Guerre Medal with a palm Oak Leaf Cluster for his heroic service to deliver 12 important messages in Verdun. He died at Fort Monmouth, New Jersey, on June 13, 1919 from the wounds he received in battle and was later inducted into the Racing Pigeon Hall of Fame in 1931.

Cher Ami was later mounted by a taxidermist and enshrined in the Smithsonian Institution in Washington, DC.

Sergeant Stubby

Sergeant Stubby (1916 or 1917-March 16, 1926) was the most decorated war dog of World War I and the only dog to be promoted to sergeant through combat.

Stubby served with the 102nd Infantry, 26th Yankee Division in the trenches in France for 18 months and participated in four offensives and 17 battles. He entered combat on February 5, 1918 at Chemin des Dames, north of Soissons and was under constant fire, day and night, for over a month. In April 1918 during a raid to take Seicheprey, Stubby was wounded in the foreleg by retreating Germans as they threw hand grenades. He was sent to the rear for convalescence and as he had done on the front, was able to improve morale. When he recovered from his wounds, Stubby returned to the trenches.

After Stubby had been gassed in an enemy gas attack, Stubby learned to warn his unit of poison gas attacks because he was so sensitive to the gas. He was able to locate wounded soldiers in no man's land because he could hear the whine of incoming artillery shells before humans could. He became very adept to let his units know when to take cover.

Stubby was solely responsible for the capture of a German spy in the Argonne after he detected him as he tried to map out his unit's trench lines. He attacked the German and continued to attack and bite him until soldiers came to investigate and took control.

At the end of the war Stubby was smuggled back home by his unit where he led the American troops in a pass and review parade and later visited President Woodrow Wilson. He visited the White House twice and met Presidents Harding and Coolidge. Stubby was awarded many medals for his heroism that included a medal from the Humane Society which was presented by General John Pershing, the Commanding General of the United States Armies. Found as an abandon stray and died an American hero. Stubby died in 1926 and is enshrined at the Smithsonian Institute.

World War II

World War II produced more heroes than any other war, or so it seemed. The war really did not have more heroes than any other war, but because of the new technology of the time, new medicines, medical procedures, advanced war planning and tactics made the public more aware of what was happening in the war, therefore making any heroic feat or situation stand out. Of course the media resources were much more advanced in WWII which made the current news of the war available to the public on a daily basis, more so than in previous wars.

The involvement of the United States in the war brought on a strong interest from the public of what happened on a daily basis. Daily reports from radio and the newspapers kept people informed of current events. Before long the public felt they should be more involved in the war.

World War II innovation brought new inventions and advanced technology at the highest point at any other time in history. Although heroic feats by an individual usually produced a hero, so did these advancements.

War Time Inventions

During the war many inventions and further advancements in technology to improve products and machinery already being used surfaced almost every day. Inventions were presented to the government by individuals who felt their invention would help the war effort and eventually bring them wealth in the process.

Hundreds, if not thousands, of inventions were submitted to the government for approval. Weapons and war machinery was the government's first concern. The government knew that the United States had to stay ahead of its enemies in weapons technology and without new weapons and improvements this would not happen.

It was amazing that during the war the government was mainly concerned to find new ways to kill the enemy, but did not make much advancement in medicine or ways to medically treat our troops. One of the most important, if not the most important, inventions in the war was Penicillin. Penicillin was actually invented in 1928, sort of. Scientists had an idea that it prevented infection but were unsure of what diseases or infections it would cure. In 1942 with new advancements in the laboratories scientists found that Penicillin would cure many different infections.

Penicillin antibiotics proved significant because they were the first drugs that were effective against many previously serious diseases and infections. There were repeated attempts to find the most effective and highest quality Penicillin. It was not until the discovery of a moldy cantaloupe found in a market in 1943 that led to the highest quality Penicillin. Discovery of the cantaloupe and the results of fermentation research allowed the United States to produce 2.3 million doses in time for the invasion of Normandy in the spring of 1944.

By June 1945 over 646 billion units per year were produced as a result of the war and the War Production Board. Penicillin made a major difference in the number of deaths and amputations caused by infected wounds which saved an estimated 15 percent of lives among Allied Forces.

The Higgins Boat

The Higgins Boat was a landing craft that was extensively used in amphibious landings in WWII. More than 23,000 were built by Higgins Industries and other companies.

Typically constructed of plywood, this shallow draft barge-style boat could carry a platoon-sized complement of 36 men to shore at 9 knots. To enter the boat men climbed down the side of a troop transport ship via a cargo net into the boat.

The first Higgins Boats, though mechanically sound, were poorly designed. Unloading of equipment and troops it had to be done over the side which made it extremely difficult and certain harm or death to disembarking troops. This was its main drawback. Modifications were made and a full-width ramp was added to the bow which made it much easier to unload and disembark troops.

Regardless of the modifications, troops were still susceptible to enemy gunfire and many were killed or wounded in these amphibious landings.

The Higgins Boat, regardless of the obvious drawback, provided the opportunity to place troops and equipment on the beaches quickly and effectively.

The Higgins Boat has had a long and successful career. These boats participated in more than 15 major amphibious landings during WWII and were successful in the Korean War and later used to patrol the rivers of Vietnam.

Liberty Ships

Though there were many innovations in WWII, the Liberty Ship was one that far surpassed manufacturing expectations.

Liberty Ships were cargo ships that were built in the United States during WWII. Though British in conception, they were adapted by the U.S. They were cheap and quick to build and came to symbolize U.S. wartime industrial output. The ship was based on ships ordered by Britain to replace ships torpedoed by German U-boats. They were purchased for the U.S. fleet and for lend-lease provision to Britain.

Eighteen American shipyards built 2751 Liberty Ships between 1941 and 1945.

The production of these ships mirrored, on a much larger scale, the manufacture of a ship similar in types used in WWI. The immense effort to build Liberty Ships, the sheer number of ships built and the fact that some of the ships survived far longer than the original design life of five years, proved that even though mass produced quickly for a nominal amount of money, some withstood the war to continue on in service to the U.S. and many other countries after the war.

Liberty Ships were constructed of sections that were welded together. This technique used welding instead of riveting. Riveted ships took several months to construct. The work force was newly trained and no one had previously built welded ships. As America entered the war, the shipbuilding yards employed women to replace the men who enlisted in the Armed Forces.

The first ships required about 230 days to build but the average dropped to 42 days. The ships were made assembly-style from prefabricated sections. In 1943 three ships were completed daily.

The pressure on the shipyards to produce these ships in mass, in a very short time and for little money did not come without problems. Even though satisfactory materials were used and not the blame for situations that occurred, manufacturing techniques were to blame. Suspicion fell on the shipyards, which in great haste, had often used inexperienced workers and the new welding techniques to produce large numbers of ships.

Early Liberty Ships suffered hull and deck cracks and a few were lost to such structural defects. There were 1500 instances of significant brittle fractures. Twelve ships, that included three of the 2710 Liberties built, broke in half without warning.

Investigations revealed that sub-standard steel was susceptible to the cold temperatures of the Atlantic Ocean. This caused the steel to become brittle and therefore caused cracks in the hull and decking. Severe storms at sea and grossly overloaded cargo-holds of the ships greatly increased the danger of damage to the ship and placed the ships in extreme danger to break up and sink. Eventually reinforcements were applied to the ships to strengthen and arrest the crack problems.

More than 2400 Liberty Ships survived WWII. Of these, 835 made up the postwar cargo fleet. Foreign entrepreneurs bought 624 of them. Some were lost after the war to Naval Mines that were inadequately cleared. In 1960 three Liberties were reactivated and used for technical research ships. The last remaining active Liberty ship was still sailing in 1994.

The Liberty Ship had a long, illustrious career serving the military and many countries after the war.

Operation Pluto

Operation Pluto (Pipe Lines Under The Ocean) was the brainchild of Arthur Hartley, chief engineer with the Anglo-Iranian Oil Company. Operation Pluto was a WWII operation by British scientists, oil companies and the Armed Forces to construct undersea oil pipelines under the English Channel between England and France.

Allied Forces on the European Continent required a tremendous amount of fuel. Pipelines were considered necessary to relieve dependence on oil tankers that could be slowed by bad weather, were susceptible to German submarines and were needed in the Pacific War.

There was a large amount of study and investigation that went into the correct type of material to be used for the piping that would carry the fuel nearly thirty miles under the ocean channel to strategic pumping stations. These were scattered on both sides of the channel and were spread out over seventy miles.

Within a year full-scale production of a two-inch pipeline was started on 14 August 1942. Six weeks later a thirty-mile length of pipeline was laid in a smaller channel to test its efficiency. Bad weather and rough seas hampered the project at first but eventually the line was placed between two landmasses and connected to test pumping stations. The rehearsal was a success, so much so that a three-inch diameter pipe rather than the two-inch was considered. This reduced the number of pipelines needed to pump the planned volume of fuel across the channel.

Full scale testing of a 45-mile pipeline proved successful and the first line to France was laid over a 70-mile stretch across the channel. The pipeline was so successful that an additional two-inch line was added. As the war moved closer to Germany 17 other pipelines were laid in different locations.

The Pluto pipelines were linked to pump stations on the English coast that were housed in various inconspicuous buildings which included cottages and garages. Though uninhabited these were intended to cloak the real purpose of the buildings. In England the Pluto pipelines were supplied by a 1000-mile network of pipelines to transport fuel from ports in England to France.

On January 1945, 300 long tons of fuel was pumped to France each day, which increased to 3000 long tons per day in March and eventually to 4000 long tons (almost 1,000,000 Imperial gallons) per day.

In total over 172,000,000 Imperial gallons of gasoline had been pumped to Allied Forces in Europe by VE Day. This provided a critical supply of fuel until a more permanent arrangement was made.

Arthur Hartley's innovation came at a time that was crucial to the war in Europe and to think of him as any less a hero than anyone else would be wrong.

Many hundreds, if not thousands, of ideas, inventions and

innovations came to be during World War II. Every one of them helped in their own certain way in the war. These inventions and the people who brought them to life are true heroes in every sense of the word.

The Ladies Of War

I could not even think of writing this book without acknowledging the only group of people whose steadfast, selfless devotion to the war effort was anything else but phenomenal.

The following women were instrumental, each in their own way, to help the war effort. Even though they were not nearly recognized enough for what they did, they have gone down in history to be one of the major factors that kept the war on track, whether in war or at home.

Rosie The Riveter

Rosie the Riveter is a cultural icon in the United States. She represented the American women who worked in factories during World War II. Many of the women worked in the factories that produced munitions and war supplies. These women sometimes took on entirely new jobs to replace the male workers who were in the military. Rosie the Riveter is commonly used as a symbol of feminism and women's economic power.

Although Rosie the Riveter took on male-dominated trades, women were expected to return to their everyday housework once the men returned from the war.

Rosie the Riveter inspired a social movement that increased the number of working American women to 20 million by 1944, a 57 percent increase from 1940. Although the image of Rosie the Riveter reflected the industrial work of welders and riveters, the majority of women filled non-factory positions in every sector of the economy. What unified these women and their experiences was that they proved to themselves, and the country, that they could do a "man's job" and could do it well.

The image most closely associated with Rosie the Riveter is J. Howard Miller's famous poster designed for Westinghouse in 1942 titled "We Can Do It!" It was modeled after a Michigan factory worker, Geraldine Doyle. This image was not actually intended to be Rosie the Riveter because Rosie the Riveter was a fictional character.

Fictional character or not, Rosie the Riveters of World War II stepped up and kept the nation in the war effort.

Martha Gellhorn

Martha Gellhorn was an American novelist, travel writer and journalist. She was considered to be one of the greatest war correspondents of the 20th century. She reported on virtually every major world conflict that took place in her 60-year career. Gellhorn was also the third wife of American novelist Ernest Hemingway from 1940 to 1945. At the age of 100, ill and almost completely blind, she committed suicide. It is ironic the same fate as her husband many years earlier.

Gellhorn first met Hemingway on a 1936 Christmas family trip to Key West, Florida. They agreed to travel together in Spain to cover the Spanish Civil War where Gellhorn was hired to report for Collier's Magazine Weekly. The couple celebrated Christmas of 1937 together in Barcelona, Spain. Later from Germany she reported on the rise of Adolf Hitler and in 1938 was in Czechoslovakia. After the outbreak of World War II she described these events in a novel. She later reported the war from Finland, Hong Kong, Burma, Singapore and Britain. Since she lacked official press credentials to witness the D-Day landings, she impersonated a stretcher-bearer and later recalled, "I followed the war wherever I could reach it." She was one of the first journalists to report from Dachau concentration camp after it was liberated.

Martha Raye

Of all the women who gave their all for the war effort, the troops and for all concerned, there has never been or will never be, a more gracious lady such as Martha Raye. Her love and passion for her compatriots, especially the soldiers, was nothing other than amazing.

Martha's career started in the 1930s. Even though she was multi-talented in the entertainment field, her real passion was to entertain the American troops.

When World War II broke out she joined the USO with the sole intention to use her talents to entertain the troops everywhere.

She traveled extensively to entertain the American troops of World War II, the Korean War and the Vietnam War even though she had a lifelong fear of flying.

In October 1966 she went to Soc Trang, Vietnam to entertain the troops at the base of the 121st Aviation Company, the Soc Trang Tigers, the gunship platoon, The Vikings and the 336th Aviation Company. Shortly after her arrival, both units were called out on a mission to extract supposed POWs from an area nearby. Raye decided to hold her troupe of entertainers there until the mission was completed so that all of the servicemen could watch her show.

A pilot flying a "Huey Slick" helicopter that was carrying troops, had his aircraft severely damaged by enemy small arms fire to the extent that he had to return to the Soc Trang base. Martha had been informed that a gunship had been shot down in a combat area. Additional efforts were made to extract the crew and word was received at that time of the death of the pilot. Raye stated that she and her troupe would remain until everyone returned from the mission. While the servicemen waited, Raye played poker with them and helped to keep everyone's spirits up.

When the mission was completed, it resulted in the loss of a helicopter gunship, a Viking pilot and a major who was in command of the Vikings was wounded when the ship went down. He was flying pilot position but was not in control of the ship when the command pilot, a warrant officer, was shot. When he and the remaining crewmen returned to Soc Trang, Raye volunteered to assist the doctor in treatment of the wounded flyer. When all had been completed, Raye waited until everyone was available and then put on her show.

There are many stories such as these about Martha and I thought that it would be fitting to include my own personal experience with this fine woman.

Our company ("Lima" Company 3rd Battalion, 7th Marines) had just completed a two-week search and destroy mission along the DMZ of Vietnam and were pulled back into a base outside of Hue to re-supply and re-group.

While at the base we were treated to a small USO show. There wasn't a large group of entertainers to put on an elaborate show for us, just Martha Raye and her two musicians. Not everyone in the unit attended the show so the crowd was rather small. No more than 40 troops attended the show in the base mess hall. One of the musicians came out onto the makeshift stage and announced Martha. She stepped out from behind a partition as the troops clapped and she automatically went into her song and dance routine while her accompanist banged away on the piano. When she finished her song, she blew a kiss to all of us and started in on a repertoire of jokes. After her jokes she addressed the group. She was polite and gracious and made us forget where and why we were in a foreign country. When she talked to us she made you feel as if she was talking to just you. A joke here a song there and she had all of us in the palm of her hand.

Someone in the crowd held up a fifth of booze, she pointed at him and held her other hand up in the air to silence the crowd, "Is that an offer?" The crowd went crazy. He nodded and pushed the bottle out further toward her. She stepped from the stage, took the bottle in her hand, stepped back slightly, climbed back onto the stage, put the bottle to her lips, tipped her head back and took a long pull from the bottle. She lowered the bottle from her lips and made a face like a teenager who had just experienced his first taste of liquor and pushed the bottle back to the Marine. She said "Hold on to this, we'll finish it later, I got a show to do." Again, the crowd went wild. He accepted the bottle and sat down.

She finished her show and exited the stage to a thunderous applause. For an hour this fine lady made us forget why we were there. She then went to the infirmary to visit with the wounded that could not make it to the show and while there she gave blood. She left the next day and we all had a good feeling.

During the Vietnam War she was made an honorary Green Beret because she visited United States Special Forces in Vietnam without fanfare. She also helped when things got bad in Special Forces A-Camps. As a result, she came to be known affectionately by the Green Beret as "Colonel Maggie."

On November 2, 1993 Martha Raye was awarded the Presidential Medal of Freedom by President Bill Clinton for her service to her country.

In appreciation of her work with the USO in World War II and subsequent wars, special consideration was given to bury her in Arlington National Cemetery upon her death. However at her request, she was ultimately buried with full military honors at Fort Bragg, North Carolina in the Post Cemetery. She was an honorary Colonel in the Marines and an honorary Lieutenant Colonel in the US Army.

Colonel Maggie is no longer with us but thousands of men and women who hearts she touched will remember her. She began to entertain in 1942 and traveled all over the world to be with them. For nine years she went to Vietnam, and sometimes stayed as long as six months. Not only did she perform on stage but also when things got rough she filled in as a nurse and often went hours without a break.

For those of us who knew her or were entertained by her we consider her as much, if not more so, a part of the Armed Forces. She wore the uniform proudly, she wore the uniform in the trenches, she wore the uniform in the field instead of on stage and she deserves to be remembered for those fifty-plus years of unique and totally dedicated military service.

Josephine Baker

Josephine Baker was an American dancer, singer and actress who found fame in her homeland of France.

Baker was the first African American female to star in a major motion picture, to integrate an American concert hall and to become a world-famous entertainer. She was the first American born woman to receive the French military honor, the Croix de Guerre.

In the early 1930s Baker performed on stage and in movies, but was not as popular in America as she was in France. In 1937 Baker returned to France where she married a Frenchman and denounced her American citizenship to become a French citizen.

Baker's affection for France was so great that when World War II broke out, she volunteered to spy for her adopted country. Baker's agent's brother approached Baker to ask her work for the French Government as an "honorable correspondent". If she agreed it was because she was against the Nazi stand on race not because she was black, but also because her husband was Jewish. Her café society fame enabled her to rub shoulders with those in-the-know from high-ranking Japanese officials to Italian bureaucrats and she was to report what she heard. She attended parties at the Italian Embassy and gathered information without any suspicion.

Baker helped the war effort in other ways also. She sent Christmas presents to French soldiers and entertained them when she was able. Baker left Paris and went to her home in the south of France when the Germans invaded. She had Belgian refugees and others living with her who were eager to help the Free French effort led from England by Charles de Gaulle. As an entertainer Baker had an excuse to move around

Europe and visit neutral nations like Portugal and then return to France. Baker assisted the French Resistance by smuggling secrets written in invisible ink on her sheet music.

Baker helped mount a production in Marseille to give herself and her like-mined friends a reason for being there. She helped a lot of people who were in danger from the Nazis to get visas and passports to leave France. Later in 1941 she and her entourage went to the French colonies in North Africa. The stated reason was Baker's health, but the real reason was to continue to help the Resistance. From a base in Morocco she made tours of Spain and pinned notes with the information she gathered inside her underwear. Later she would perform at Buchenwald Concentration Camp for the liberated inmates who were too frail to be moved.

In the 1950's she was very involved in the Civil Rights Movement. She adopted 12 multi-ethnic orphans. She refused to perform before segregated audiences in the United States. Her insistence on mixed audiences helped to integrate shows in Las Vegas.

Baker worked with the NAACP. In 1963 she spoke at the March on Washington at the side Martin Luther King. After King's assassination, his widow Coretta Scott King, approached Baker and asked her to take over her husband's place as leader on the American Civil Rights Movement. After many days to think it over, Baker declined, and said her children were, "too young to lose their mother".

Nancy Wake

Nancy Grace Augusta Wake, nicknamed "The White Mouse", served as a British agent in the latter part of World War II. She became a leading figure in the Maquis groups of the French Resistance and was one of the Allies most decorated servicewomen of the war.

In 1937 she met wealthy industrialist, Henri Fiocca, whom she married in 1939. She was living in Marseille, France when

Germany invaded. After the fall of France in 1940 she became a courier for the French Resistance and later joined the escape network to help refugees leave France. In reference to her ability to evade capture, the Gestapo called her "The White Mouse". The Resistance had to be very careful with her missions. Her life was in constant danger because the Gestapo tapped her phone and intercepted her mail.

By 1943 she was the Gestapo's most wanted person with a five million Franc price on her head. When the network was betrayed that same year, she decided to flee France. Her husband stayed behind where he was later captured, tortured and executed by the Gestapo.

On the night of April 29-30, 1944 she was parachuted into the Auvergne to become a liaison between London and the local Maquis group that were headed by Captain Tardivat. When he discovered her tangled in a tree Captain Tarivat greeted her by remarking, "I hope all the trees in France bear such beautiful fruit this year," to which she replied, "Don't give me that French shit."

Her duties included allocation of arms and equipment that were parachuted in and handle the group's finances. She became instrumental in recruiting more members to help make the Marquis groups into a formidable force of roughly 7500 strong. She also led attacks on German installations and the local Gestapo HQ in Montlucon.

From April 1944 to the liberation of France, her 7000 Maquisards fought 22,000 SS soldiers and caused 1400 casualties, while they only sustained100 themselves. Her French companions, especially Captain Tardivat, praised her fighting spirit that amply was demonstrated when she killed an SS sentry with her bare hands to not allow him to raise the alarm during a raid. On another occasion, Wake rode a bicycle for more than 500 miles through several German checkpoints to replace codes her wireless operator had been forced to destroy in a German raid.

Wake, along with two American officers, took command of

a section whose leader had been killed in an attack on another Marquis group. She directed the use of suppressive fire which facilitated the group's withdrawal without further losses.

As a heroine of World War II, Wake earned more than 10 awards from multiple countries that included the Medal of Freedom from the United States.

Soviet Women In World War II

Soviet women bore their share of burden in World War II (locally known as the Great Patriotic War). While most worked double shifts in industry, transport, agriculture and other civilian roles to free up enlisted men to fight and increase military production, a sizable number of women took up arms.

More than 800,000 women served in the Soviet Armed Forces in the war. Nearly 200,000 were decorated and 89 eventually received the Soviet Union's highest award, the Hero of the Soviet Union. They served as pilots, snipers, machine gunners, tank crew members, partisans, as well as in auxiliary roles.

At first when Germany attacked the Soviet Union on June 22, 1941, thousands of women who volunteered were turned away. Two factors changed the attitudes and ensured a greater role for women who wanted to fight; the losses to the Germans after their initial success in 1941 and the efforts of determined women. In the early stages of the war the fastest route to advancement in the military for women was service in the medical and auxiliary units.

Women Pilots

For Soviet women aviators Maria Raskova, a famous Russian aviator, was instrumental to this change of women who participated in war and was often referred to as the 'Russian Amelia Earhart'. Raskova was the first woman to become both a pilot and a navigator in the 1930s. She was the first women to become a navigator in the Soviet Air Force in 1933. A year later she became a teacher at the Zhukovskii Air Academy, also a first for a woman.

When World War II broke out numerous women who had trained as pilots, immediately volunteered. While there were no formal restrictions on women to serve in combat roles, their applications tended to be blocked and run through red tape, etc. for as long as possible in order to discourage them to see combat. Raskova is credited with using her personal connections with Joseph Stalin to convince the military to form three combat regiments for women. Not only would the women be pilots but also the support staff and engineers for these regiments. The Soviet Union was the first nation to allow women pilots to fly in combat missions. These regiments flew a combined total of more than 30,000 combat sorties that produced at least 30 Heroes of the Soviet Union and included at least two fight aces. Lydia Litvyak and Katya Budanova were assigned to 437th IAP in the fighting over Stalingrad and became the world's only two female fight aces (with 12 and 11 victories respectively).

Land Forces

The Soviet Union deployed women snipers extensively that included Nina Alexeyevna Lobkovskaya and Ukraininian Lyudmila Pavlichenko, who killed over 300 German soldiers. The Soviets found that women were better suited for sniper duties than men because good snipers are patient, deliberate, have a high level of aerobic conditioning and normally avoid hand-to-hand combat.

Women served as machine gunners, tank drivers, medics, communication personnel and political officers. Women crewed the majority of the anti-aircraft batteries employed in Stalingrad.

Stalin allowed a plan which would replace men with women in second lines of defense, such as anti-aircraft guns and medical aid in response to the high casualties suffered by male soldiers. These provided gateways through which women could gradually become involved in combat and demonstrate their capabilities. For example, women comprised 43 percent of physicians who were often required to carry rifles as they retrieved men from the firing zones. Through opportunities like this women gradually gained credibility on the battlefield and eventually numbered 500,000 at any given time toward the end of the war.

Women Air Force Service Pilots (WASP)

The Women Air Force Service Pilots (WASP) and their predecessor groups, the Women's Flying Training Detachment (WFTD) and the Women's Auxilary Ferring Squadron (WAFS), were pioneer organizations of civilian female pilots employed to fly military aircraft under the direction of the United States Army Air Forces. In August of 1943 the WAFS and the WATD were combined to create the paramilitary WASP organization. The female pilots of the WASP would end up with 1074 members of which each freed a male pilot for combat service and duties. The WASP flew over 60 million miles in all in every type of military aircraft.

Twenty-five thousand women applied to join the WASP but only 1830 were accepted and took the oath. Of those only 1074 women passed the training and joined.

After training the WASPs were stationed at 120 bases across the U.S. to assume numerous flight-related missions. They flew sixty million miles of operational flights from aircraft factories to ports of embarkation and military training bases, towed targets for live anti-aircraft artillery practice, simulated strafing missions and transported cargo. Almost every aircraft flown by the USAAF during World War II was also flown at some point by women in these roles. In addition a few exceptionally qualified women were allowed to test rocket-propelled planes, pilot jet-propelled planes and work with radar-controlled targets. The WASP delivered 12,650 aircraft of 78 different types between September 1942 and December 1944.

Thirty-eight WASP fliers lost their lives in accidents while serving during the war; eleven in training and twenty-seven on active duty. A fallen WASP was sent home at family expense without traditional military honors or note of heroism because they were not considered to be in the military under the existing guidelines. The Army would not even allow the U.S. flag to be put on fallen WASP coffins.

I feel it fitting to include my personal recognition of a WASP pilot I met in 2009. I was lucky enough to be introduced to Lizabeth Ann Morgan Hazzard while I was wintering in southern Texas. Ann graduated from Texas Women's University and enrolled in the WASP program. She trained at Avenger Field in Sweetwater, Texas after graduating from WASP class 43-W-7. After training she was stationed at Mather Army Air Base and March Army Base. In her years as a WASP she primariliy flew AT-6s, B25s and B17s to numerous locations across the U.S.

In 1977 WASPs were granted veterans status. In 2009 the remaining 300 WASPs were flown to Washington DC where President Barack Obama awarded them the Congressional Gold Medal for their service to their country. At the age of 85, Lizabeth Ann Morgan Hazzard had been recognized for her heroic feats during World War II. A fitting long overdue tribute to Ann and the WASP's of WWII.

Air Spotters of WW II

In World War II the government established a network of civilian airplane spotters whose duty it was to thwart any Axis sneak attacks such as Pearl Harbor. From lonely observation posts throughout coastal defense areas, the Army Air Forces Ground Observer Corps, that numbered about 1,500,000 volunteers, maintained a vigilant and continuous watch on the skies over America to see that no hostile planes approached land unnoticed or undetected. This volunteer force was set up by the Fighter Command, with the assistance of the American Legion, the OCD and other agencies to establish a new pattern of scouting. It remained in place until advanced technology in radar made them no longer necessary.

The town of Kent, Connecticut is credited to be the first such observation post in America that was established on December 8, 1941. It became the model on which the entire program was based. The observation posts were erected on a prominent hill or rise that offered an unobstructed view.

Everyone from high school students to senior citizens made up the force that manned these observation posts around the clock. Spotters usually worked in pairs for two to four hours a week and relayed observations by special telephone to the Army Information Center. On fair weather days the observers relied on their eyes to identify aircraft. On days that were cloudy or overcast and at night they used theirs ears to interpret the situation, which not always turned out as accurate information.

There were decks of cards that featured all sorts of aircraft, both Allied and Axis, that were studied until they were committed to memory. Forms were supplied on 3 x 5 cards that had these categories: number of aircraft (one, few, many), type of aircraft (single engine, multi-engine), altitude of aircraft (very low, low, high, very high), were aircraft seen or heard? (check one), your observation post code name, direction of aircraft from post (N, NE, etc.), distance from observation post (estimated in miles), and aircraft headed toward (give point direction of compass).The observer would pick up the phone and report their observation to the command before the card was filled out. Of course this plane would be reported by one station after another and the Army was able to track any and all aircraft anywhere within the covered area.

Only one time during the course of the war did a German aircraft fly into American airspace and that was near the end of the war when an Army Air Force crew flew a captured German plane to Florida. Of course the military knew about this, but the civilian observers were kept in the dark just to test their proficiency. A spotter sent an emergency message that not only identified it as a German aircraft, but also the correct make and model before the plane crossed from water to land.

As amazing as these observation posts were in the U.S., there were thousands upon thousands of them across the globe. Allied countries took advantage of this network of spotters and placed them all over the world. Posts were placed on the coastal areas of nearly every country and most continents of the world. The least known of these observation posts were those individuals dropped off on small uninhabited islands in the South Pacific. In the dead of night a ship would sneak in and drop an individual along with enough supplies to carry them over for a few months. In addition to the supplies, sophisticated radio equipment, a supply of batteries and a small hand cranked generator to be used as an alternative source of power for the radio were supplied. Most times there were no weapons issued because these individuals were to be invisible and if discovered by the enemy they were to disappear.

These island observers were hidden from sight in a small camouflaged hut nearly invisible to any aircraft that may pass over. The radio only had a certain range to transmit messages so a network of locations would pass on the information from one island to another until the information reached the command.

If the enemy happened to land on an island occupied by an observer, the observer would have an unobstructed view of most of the island and would have enough time to gather up all equipment and supplies and retreat to an alternate hiding place where the enemy could not find them.

These observers were an invaluable source of vital information and were a total volunteer network of individuals.

The Merchant Seaman of World War II

The first combat fatality of WW II was a merchant seaman when a German U-boat sank the liner Athenia off the coast of

Ireland. There was heavy loss of life to crew and passengers. The last person to die after the cessation of hostilities with Germany also was a merchant seaman when his ship was torpedoed in the North Atlantic three days after the war had ended.

It is not generally known that all hospital ships, royal fleet auxiliaries, armed merchant cruisers and most of the Woolworth-style carriers of WW II were crewed by merchant seaman. They sailed under the Red Ensign, Blue Ensign and the White Ensign. The merchant seaman did their job without fuss or fanfare and very often went largely unnoticed whether they were on passenger liners, troopships, ocean going tugs, cargo ships, cross channel ferries or fishing boats.

Merchant ships were not built for war and merchant seaman were not trained for war, whereas naval vessels were specifically designed to absorb damage from enemy attack and their crews were highly trained in damage control and gunnery.

The horrors of the North Atlantic convoys, which in most instances were completely unable to defend themselves, were at the mercy of a ruthless enemy for up to a month at a time day and night. Between 1939 and 1945, in the so called Battle of the Atlantic, nearly 3000 Allied and neutral merchant ships were sunk by enemy action with a heavy loss of life. Nightmare trips to the extreme North Atlantic ports suffered the rigors of an Arctic winter while under continuous enemy attack. These trips were endured by crews of the merchant vessels to ensure that essential war equipment, supplies and foodstuffs got through to our Russian Allies.

In August 1942 one of the greatest sea battles of World War II, Operation Pedestal, was fought by 14 merchant ships and a large fleet of naval vessels. Throughout five days of continuous warfare from air and sea the fleet endeavoured to get essential supplies to Malta. Not many people have heard of this operation because it was considered just another convoy of that time. There were more merchant seamen lost in that five days of fighting than Australia lost in the whole of the Vietnam War.

Generally a convoy of 45 ships had an escort of about four or five corvettes and maybe one destroyer. Due to the importance of this convoy of 14 merchant ships to Malta there was an escort of two battleships, five aircraft carriers, seven cruisers, 22 destroyers, seven submarines and 14 other naval vessels. Yet despite this level of support, at the end of the five days only five merchant ships had survived and three of them were badly damaged.

Wherever the war was waged, merchant ships were there taking troops and essential supplies to the heart of the action. The traversing of supply lines by these vessels and their crews was an extremely dangerous occupation. The whole maritime was their battleground. These ships and their crews were at war from the time they left port not knowing when they might be attacked, blown up or disabled in some fashion. Here are the gruesome statistics of what the merchant seamen endured during the war.

1. 4996 British and Allied ships lost
2. 62,933 British and Allied merchant seamen killed in action
3. 4000 wounded
4. 5000 merchant seamen taken as prisoners of war

Most people have heard about the merchant seamen and what horrific situations they endured by the enemy, but most do not know what they endured under the command of their own countries.

When a merchant ship was sunk the seamen's pay stopped on the day of the sinking. He did not receive any more pay until he joined another ship. The seaman was given 30 days survivor's leave dated from the day his ship was sunk. This leave was unpaid. It meant that he didn't have to report back to the pool for thirty days. If he spent 10 or 15 days in a life raft or in a lifeboat, that counted as survivor's leave.

There were many merchant seamen who joined the Navy because it was extremely short of experienced seamen. These men were in Naval uniforms on Naval vessels under the White

Ensign, with Naval officers and subject to Naval discipline. They received Naval rates of pay. At the end of the war they were not allowed to claim any compensation or any benefits because they were discharged as merchant seamen.

Women POW's of World War II

Two days after the bombing of Pearl Harbor five Navy nurses on Guam were taken prisoner by the Japanese: Lieutenants (JG) Leona Jackson, Lorraine Christensen, Virginia Forgerty, Doris Yetter and their commander Chief Nurse, Marion Olds. Later in 1942 their captors transported them to Japan. They were held three months in Zentsuji Prison on Shikoku Island and were moved to Eastern Lodge in Kobe. They were repatriated in August of 1942.

Clara Gordon Main, a stewardess on the SS President Harrison, was captured by the Japanese on December 7, 1941 while rescuing U.S. Marines from China. She was among the first American Prisoners of War.

Agnes Newton Keith was imprisoned in several Japanese prison camps from 1941 to the very end of the war.

Many more women were imprisoned during World War II and used in many capacities by their enemies. Some were used in manufacturing capacities while most were put to work as nurses to assist doctors in treatment of enemy sick and wounded. These women prisoners were treated quite well even though their living conditions were atrocious.

Some nurses were required to establish make shift infirmaries in some of these prison camps, even though they had virtually no medicine or supplies to nurse or treat the sick. In spite of the work and good intentions by the nurses, the survival rate of the sick and wounded prisoners was less than 20 percent.

World War II Celebrities

Many famous celebrities served in World War II and in other wars. I thought that it would be an interesting subject to add to this book. Many on screen and television celebrities performed acts of heroism and endured horrific situations while in the service. Some of these will surprise you.

Lee Marvin

Marvin was expelled from many schools before attending St. Leo Preparatory College in Florida. He became bored and quit to enter the United States Marine Corps and served as a scout sniper in the 4th Marine Division. He was wounded in action in the Battle of Saipan, where most of his platoon were killed. Marvin's wound in the buttocks which severed his sciatic nerve from machine gun fire. He was awarded the Purple Heart and was given a medical discharge with the rank of Private First Class. Contrary to rumors, Marvin did not serve with producer and actor Bob Keeshan (Captain Kangaroo) during WWII.

Don Adams

Don Adams, of Get Smart fame, served with the 3rd Marines in 1941. Adams was sent to the Battle of Guadal Canal where he was the only survivor of his platoon.

Gene Autry

Gene Autry served as a C-47 Skytrain pilot in the USAAF. He flew many dangerous missions over the Himalayas between Burma and China to deliver vital supplies to many locations.

Eddie Albert

Eddie Albert, of Green Acres fame, toured Mexico as a clown and high-wire artist with the Escalante Circus. He secretly worked for the U.S. Army in Intelligence and photographed German U-Boats in Mexican harbors. On September 9, 1942 Albert enlisted in the U.S. Navy and was discharged in 1943 to accept an appointment as a lieutenant in the U.S. Naval Reserve. He was awarded the Bronze Star with Combat "V" for his actions in the invasion of Tarawa in November 1943. As the pilot of a U.S. Coast Guard landing craft, he rescued 47 Marines who were stranded offshore and supervised the rescue of 30 others, all while under heavy enemy machine-gun fire.

James Arness

James Arness, of Gunsmoke fame, wanted to be a Naval fighter pilot but feared his poor eyesight would bar him. His height of 6 feet 7 inches ended his hopes, since 6 feet 2 inches was the limit for aviators. Instead he was called for the Army and reported to Fort Snelling, Minnesota in March of 1943. Arness served as a rifleman with the U.S. 3rd Infantry Division and was severely wounded during Operation Shingle at Anzio, Italy. Arness landed on Anzio Beachhead on January 22, 1944 as a rifleman. Due to his height he was the first to be ordered off the landing craft to determine the depth of the water and it came up to his waist. On January 29, 1945 after several surgeries, Arness was honorably discharged. His wounds bothered him in his later years and he suffered from chronic leg pain. If you noticed when watching Gunsmoke, he walked with a slight limp.

Mel Brooks

Participated as a rifleman in the battle of the Bulge and was involved in many other key operations.

Charles Bronson

In 1943 Bronson enlisted in the U.S.Army Air Force and served as an aerial gunner in the 760th Flexible Gunnery Training Squadron. In 1945 he was a B-29 Superfortress crewman with the 39th Bombardment Group based on Guam. He was awarded the Purple Heart for wounds received during his service.

Charles Durning

Durning, known for his many achievements in the film industry, is probably best known for his role as Father Hubley on Everybody Loves Raymond.

Durning served in the U.S. Army. Drafted at age 21, he was first assigned as a rifleman with the 398th Infantry Regiment. Later he served overseas with the 3rd Army Support Troops and the 384th Anti-Aircraft Artillery Battalion. Durning was awarded the Silver Star and three Purple Hearts for his valor and wounds received.

Durning participated in the Normandy Invasion of France on D-Day on June 6, 1944 and was among the first troops to land at Omaha Beach.

He was wounded by a German "S" Mine on June 15, 1944 at Les Mare des Mares, France. By June 17th he was back in England at the 217th General Hospital. Although severely wounded by shrapnel in the left and right thighs, the right hand, the frontal region of the head and the anterior left chest wall, Durning recovered quickly and was determined to be fit for duty on December 6, 1944. He arrived at the front in December 1944 to take part in the Battle of the Bulge, the

German counter-offensive through Ardinnes Forest of Belgium and Luxembourg.

After being wounded again, this time in the chest, Durning was repatriated to the U.S. He remained in Army hospitals to receive treatment for wounds until he was discharged at the rank of Private First Class on January 30, 1946.

Art Carney

Art Carney, best known as Ed Norton on the Honeymooners of the late 50's, was drafted as an Infantryman during World War II. In the Battle of Normandy he was wounded in the leg by shrapnel and walked with a limp for the rest of his life.

Rod Serling

Rod Serling, best known as host, creator and producer of Twilight Zone of the early 1960's, started his military career at Camp Toccoa, Georgia in 1943. He served in the 511th Parachute Infantry Regiment of the 11th Airborne Division.

April 25, 1944 was the day Serling had looked forward to, the day he received his overseas orders. When he saw that he was headed west through California, he knew he was headed to fight the Japanese rather than the Germans. He was disappointed because he was Jewish, he had hoped to have a hand in combating Hitler.

On May 5 the division boarded the USS Sea Pike that was headed into the Pacific and ended up in New Guinea, where they would be held in reserve for a few months.

It wasn't until November 1944 that these troops would see combat on the Philippine Island of Leyte. The 11th Airborne Division would not be used as paratroopers, however they were sent in as light infantry after the Battle of Layte Gulf to help mop up after six divisions that had gone ashore earlier. Their mission seemed simple, go from point A to point B and clean out Japanese positions as they went. In reality the terrain and lack of military intelligence proved to be just as difficult to handle as the unpredictable enemy.

For a variety of reasons Serling was transferred to the 511th demolition platoon, nicknamed the 'death squad' for its high casualty rate. As stated by Sergeant Lewis, leader of the demolition squad, "Serling screwed up somewhere along the line. Apparently he got on someone's nerves." Lewis also noted that "Serling was not cut out to be a field soldier He didn't have the wits or aggressiveness required for combat." Serling, Lewis and others were in a firefight trapped in a foxhole. As time passed and they waited for darkness, Lewis noticed that Serling had not reloaded any of his extra magazines. Another example of how Serling was a dreamer in a harsh reality was that he would go off to explore on his own, against orders and then get lost.

Serling's time in Leyte would shape his writing of his political views for the rest of his life. He witnessed death every day while in the Philippines, both at the hands of the enemy and through random events. In his future writing career, Serling would set several of his scripts in the Philippines and used the unpredictability of death as a source for much of his material.

Serling marched away from the successful mission in Layte with two wounds, including one to his kneecap, but neither was enough to keep him from combat. When General McAuthur used the paratroopers, as they were intended, on February 3, 1945 the 511th Parachute Infantry Regiment landed on Tagaytay Ridge, met up with the 188th Glider Infantry Regiment and marched into Manila. There was minimal resistance until they reached the city where Japanese vice Admiral Sanji Iwabuchi had barricaded his 17,000 troops behind a maze of traps and guns and ordered them to fight to the death. The next month Serling's unit was involved in a block-by-block battle for control of Manila.

As the troops continued to move in on Iwabuchi's strong hold Serling's regiment suffered a 50 percent casualty rate, with over 400 men killed. Serling was wounded, and three of the men he was with were killed by shrapnel from rounds fired by an anti-aircraft gun at his roving demolition team. He was sent to New Guinea to recover but soon chose to return to Manila to finish "cleaning up". Private Serling's final assignment was as part of the occupation force in Japan.

James (Jimmy) Stewart

There were many celebrities who served during World War II. Some of them were already celebrities before entering the war and some were just civilians who later became celebrities. A few such as Audey Murphy, were soldiers who because of their heroic actions became celebrities.

Jimmy Stewart was the epitome of what it was to be a soldier. Stewart obtained his Private Pilot certificate in 1935 and Commercial Pilot Certificate in 1938. He often flew cross country navigating by the railroad tracks that crossed the nation to visit his parents in Pennsylvania. By December 7, 1941 Stewart had accumulated over 400 hours of flight time.

In 1940 Stewart was drafted into the U.S. Army but was rejected because he failed the height and weight requirements for new recruits. Stewart was five pounds under the standard weight requirement. To get up to 148 pounds, he sought out the help of Metro-Goldwyn-Mayer's muscle-man and trainer Don Loomis, who was noted for his ability to add or subtract pounds in his studio gymnasium. Stewart subsequently attempted to enlist in the Army Air Corps, but still came in under the weight requirement. Although he persuaded the AAC enlistment officer to run new tests, this time he passed the weigh-in. With that result Stewart enlisted in the U. S. Army March 1941. He became the first major movie star to wear a military uniform in World War II.

Stewart enlisted as a private and then began pilot training in the US Army Air Corps. Stewart continued his military training and earned a commission as a second lieutenant in January 1942, just after the attack on Pearl Harbor brought the US directly into the conflict. He was post at Moffett field and then Mather Field as an instructor in single-and twin engine aircraft.

Stewart was concerned that his expertise and celebrity status would relegate him to instructor duties "behind lines." His fears were confirmed when he was stationed for six months at Kirkland Air Force Base in Albuquerque, New Mexico to train bombardiers. He was transferred to Hobbs AAF to become an instructor pilot for the four-engine B-17 Flying Fortress, where he trained B-17 pilots for nine months at Gowen Field in Boise, Idaho.

Still the war was moving on and Stewart realized that he was really not a part of it. For 36-year old Stewart combat duty seemed far away and unreachable and he did not have any clear plans for the future. But then a rumor that Stewart would be taken off flying status and assigned to make training films and sell war bonds called for his immediate and decisive action. What he dreaded most was the hope-shattering spectra of a dead end. Stewart appealed to his commander, a pre-war aviator, who understood the situation and reassigned him to a unit to go overseas.

August 1943 Stewart was assigned to the 445th Bombardment Group at Sioux City Air Force Base, Iowa as operations officer and the 703rd Bombardment Squadron as its commander at the rank of Captain. In December the 445th flew its B-24 Liberator bombers to RAF Tibenham, Norfolk, England and immediately began combat operations. Stewart was promoted to Major while flying missions over Germany.

In March 1944 he was transferred as Group Operations Officer to the 453rd Bombardment Group, a new B-24 unit that had experienced difficulties. As a means to inspire his new group, Stewart flew as command pilot in the lead B-24 on numerous missions deep into Nazi-occupied Europe. He did not want the public to know how many actual combat missions he flew so these missions went uncounted for at Stewart's orders. His "official" total is listed as 20 and is limited to those with the 445th.

In 1944 he twice was awarded the Distinguished Flying Cross for actions in combat and was awarded France's highest military honor, the Croix de Guerre. He also received the Air Medal with three oak leaf clusters. July 1944 after he flew 20 combat missions, Stewart was made Chief of Staff of the 2nd Combat Bombardment Wing of the Eighth Air Force and he was no longer required or expected to fly missions but he continued to do so. Before the war ended he was promoted to Colonel, one of the few Americans to rise from private to Colonel in four years.

Stewart continued to play a role in the United States Air Force Reserve after the war and achieved the rank of Brigadier General on July 23, 1959. Stewart did not often talk of his wartime service, perhaps due to his desire to be seen as a regular soldier who did his duty instead of as a celebrity.

After the war Stewart served as Air Force Reserve Commander of Dobbins Air Reserve Base in the early 1950's. In 1966 Brigadier General James Stewart flew as a non-duty observer in a B-52 on a bombing mission in the Vietnam War. At the time of his B-52 flight he refused the release of any publicity about his participation as he did not want the notoriety or the credit for doing so.

Ed McMahon

Ed McMahon, best known as Johnny Carson's sidekick, hoped to become a United States Marine Corps fighter pilot. Prior to the US entry into World War II, both the Army and the Navy required two years of college for their pilots program. McMahon enrolled into classes at Boston College. The college requirement was dropped after Pearl Harbor was attacked and McMahon immediately applied for Marine flight training. His primary flight training was in Dallas, Texas followed by fighter training in Pensacola, Florida where he also earned his carrier landing qualifications. He was a Marine flight instructor for two years and finally was ordered to the Pacific fleet in 1945.

However his orders were canceled after the atomic bomb was dropped on Hiroshima and Nagasaki which forced Japan's surrender. As an officer in the reserves, McMahon was recalled to active duty for the Korean War. This time he flew the OE-1 (the original Marine Corps designation for the Cessna O-1 Bird Dog), an unarmed single-engine spotter plane. He functioned as an artillery spotter for the Marine batteries on the ground and as a forward controller for the Navy and Marine fighter bombers. He flew a total of 85 combat missions and earned six Air Medals. After the war he stayed with the Marines as a reserve officer and retired in 1966 as a Colonel. He was later commissioned to the rank of Brigadier General in the California National Guard.

Charles Schulz

Charles Schulz, of Peanuts fame, was drafted into the U.S. Army in 1943 and served as a sergeant with the 20th Armored Division in Europe as a squad leader on a .50 caliber machine gun team. The unit saw combat only at the end of the war. Schulz stated that he only had one opportunity to fire his machine gun, but forgot to load it. Fortunately, he said, the German soldier he ran into willing surrendered. Years later he spoke freely of his wartime service.

Julia Child

Child joined the Office of Strategic Services (OSS) after she learned that she was too tall to enlist in the Women's Army Corps (WAC'S) or in the U.S.Navy WAVES. She began her career as a typist at its headquarters in Washington, but because of her education and experience soon was given a more responsible position as a top secret researcher and worked directly for the head of OSS, General William J. Donovan. As a research assistant in the Secret Intelligence Division, she typed 10,000 names on white note cards to keep track of officers. For a year she worked at the OSS Emergency Rescue Equipment Section (ERES) in Washington D.C. as a file clerk and then as an assistant to developers of a shark repellent needed to ensure that sharks would not explode ordinance used to target German U-Boats. In 1944 she was sent to Kandy, Ceylon (now Sri Lanka), where her responsibilities included to register, catalogue and channel a great volume of highly classified communications for the OSS's clandestine stations in Asia. She was later posted to China, where she received the Emblem of Meritorious Civilian Service as head of the Registry of the OSS Secretariat. As with all other OSS records, Child's OSS file was declassified in 2008 some 65 years after she performed these duties.

I could go on and on listing celebrities that served during wartime but instead I will list those celebrities that did their own part towards the war effort. Some of these will surprise you.

Richard Boone (Paladin-Have Gun-Will travel)
Robert Clary (Le Beau-Hogan's Heroes)
Jackie Coogan (Uncle Fester-Addams Family)
Alan Hale Jr. (Gilligan's Island)
Don Knotts (Andy Griffith Show)
Steve McQueen (Actor)
Burgess Meredith (The Penguin on Batman, Rocky)
Carroll O' Connor (All In The Family)
Jack Palance (Western movie actor)
Don Rickles (Comedian)
Rod Stieger (Actor)
Humphrey Bogart (Actor)
Tony Curtis (Actor)
Bea Authur (Actress/Comedienne)
Tennessee Ernie Ford (Singer)
Bob Barker (Price Is Right)
Walter Mathau (Actor)
Jack Lemmon (Actor)
Dezi Arnez (Actor/ Singer)

Ernest Borgnine (McHales Navy)
Mickey Rooney (Actor)
Burt Lancaster (Actor)
Marty Robbins (Singer)
De Forest Kelly (Bones on Startrek)
Buddy Hackett (Comedian)
Jack Parr (Tonight Show Host)
George Reeves (Superman 1950's)
Jonathan Winters (Comedian)
Harvey Korman (Carol Burnett Show)
Carl Reiner (Actor/Producer)
Dan Rowen (Laugh-In)
Paul Newman (Actor)
Denver Pyle (Dukes of Hazzard)
Mike Connors (Manix)
Larry Storch (F Troop)
Werner Kleperer (Colonel Klink, Hogan's Heroes)

Those above and hundreds more like them decided that a career in the entertainment business was not as important as serving their nation in the time of war. While some of these entertainers chose to serve in a capacity that was less than dangerous, many did serve in actual combat situations. Regardless in what capacity they served, they deserve to be known as heroes.

The Korean War

The Korean War came just six short years after World War II ended and threw the United States into another war before it could fully recover completely from the previous war. There were few new innovations / inventions during this war that could be called monumental. A good portion of the weapons and materials used in The Korean War were left over or surplus from the previous war.

There was little advancement in weaponry. Small arms such as rifles, automatic weapons such as the sub machine gun, grease gun and crew-served weapons such as machine guns, mortars and the like were all used during the Korean War and also into the Vietnam War.

Warehouses, underground storage facilities stockpiled with medical supplies, food stores and munitions left over from World War II helped in the start of the Korean War. When these stockpiles were diminished new ones were produced. When these supplies were used up it made way for some changes in quality and design.

Although these changes did not affect many of the weaponry used during the war, some of the weapons used in World War II were the main weaponry used for the Korean War. Some were also used throughout the Vietnam War.

As I have said previously in this book, not all heroes are people. Inanimate items can also be characterized as heroes, or as in this case, unheralded heroes. These items or objects have made such an impact on the outcome of wars that they need to be recognized.

The C-Ration Meal

The C-Ration, or Type C Ration, was an individual canned, pre-cooked or prepared wet ration, intended to be issued to U.S. military land forces when fresh food or packaged unprepared food to be prepared in mess halls or field kitchens was impractical or not available. Also used when a survival ration was insufficient.

The Field Ration, Type C (1938-1945), replaced the "Reserve Ration" (1917-1937) used in and beyond World War I.

Development of a replacement for the Reserve Ration was undertaken by the newly formed Quartermaster Subsistence Research and Development Laboratory in Chicago in 1938. The aim was to produce a ration that was more palatable, nutritionally balanced and had better keeping qualities.

Initially C Ration cans were marked only with paper labels, which soon fell off and made what was within a guessing game for the evening meal. Marines and U.S. Soldiers who received an unpopular menu item several nights in a row often found themselves powerless to bargain for a more palatable one.

The C Ration was, in general, not well liked by U.S. Army or Marine Forces in World War II. They found the cans heavy and cumbersome and the menu monotonous after a short period of time. There were also inevitable problems with production consistency given the large number of suppliers involved and the pressures of wartime production.

Originally intended only for infrequent use, the urgencies of combat sometimes forced supply authorities to make the C Ration the only source of sustenance for several weeks in succession.

While the initial specification was officially declared obsolete in 1945 and production of all C Rations ended in 1958, existing stockpiles of both original and revised Type C

Rations continued to be issued to troops that served in Korea and even as late as the Vietnam War. I personally ate hundreds of cans of Korean Era C Rations that were stamped 1951, 52 and 53. Since I consumed many of these 50's style C Rations, I personally found that the late 40's revised C Ration offered a good selection of menu items and was quite tasty. A person could fashion quite a good meal with a little imagination and a bottle of Tabasco sauce with what was not tasty. In Vietnam we Marines actually preferred the C Ration meal over mess hall food mainly because the C Rations consisted of an actual prepared food. It was not something in powder form where water was added to produce something that looked, but rarely, tasted like food.

Medevac Helicopters

In World War II helicopters were used for numerous light utility duties such as to scout and search for submarines. They also carried out a large medevac operation in June 1945 when helicopters airlifted at least 70 wounded soldiers from the front lines of Luzon to rear area hospitals. This marked the first time that U.S. helicopters came under concentrated enemy fire. Few helicopters, either German or American, made it into frontline service. That changed by the time the Korean War took place.

Two helicopters, the Bell 47 (designated the H-13 by the military) and the Sikorsky S-51 (designated the H-5 by the military), were the primary rescue helicopters in the Korean War. The Sikorsky 51 helicopters were pressed into service early in the war as aerial ambulances. These helicopters could carry two crew members and a wounded soldier. There was so little room in the narrow fuselage for the stretcher that the soldier's legs stuck out the side of the helicopter from the knees down.

The Bell 47/ H-13 had a two seat cockpit enclosed by a distinctive plastic bubble. The two-bladed rotor made a "chop-chop" sound, which led to the nickname "chopper" for helicopters. It became the first commercial helicopter beginning in the early 1950's. It is perhaps most famous for its extensive use in the Korean War.

Soon the H-13, which served in a light utility and observation role, was converted to a flying ambulance to ferry wounded troops from the front lines to Mobile Army Surgical Hospitals (MASH) that were often far from the front.

The H-13 normally could carry a pilot and two passengers in the bubble cockpit plus two wounded on stretchers atop the landing skids. Small plastic bubbles were fitted at the front of the stretchers to protect the men's heads. It was a cold and windy ride for the soldiers, but it meant that they would be rushed to a team of trained doctors in a well-equipped hospital. The alternative was first aid on the battlefield and a long, bumpy ride to a field hospital. However the limited carrying capacity of these small helicopters and their slow speed of 105 miles per hour while empty were clear drawbacks for military use.

Though its slow speed and carrying capacity was a drawback its maximum range of 215 miles more than made up for these problems, especially since Korea had a relatively fixed battle front during the second half of the war. MASH units could be located near the front and the helicopters did not have to fly far or for long periods of time to transport the wounded.

The United States Marine Corps adopted the H-13 and designated it the HTL-4 during the Korean War. A prototype of the helicopter was developed soon after the HTL-4 was accepted by the Marine Corps The HTL-1 was now the Marine Corps main helicopter and served the Corps well until the end of the war. In nearly two years the HTL-1 pilots of the 1st Division Aircraft Wing flew over 118,000 sorties in support of the U.N. forces. Almost 40,000 sorties were close-in support missions and Marine helicopter squadrons evacuated more than 10,000 wounded personnel and greatly increased the survival rate for wounded Marines.

Mobile Army Surgical Hospital (MASH)

In World War II the Echelon III hospital units of the field hospital were often employed in close proximity to a division clearing station where they could provide more definitive care than that available in the division. These were the predecessor to the Mobile Army Surgical Hospital (MASH) that would be employed in Korea a few years later. The lack of sufficient surgeons and nurses in the hospital units of the field hospital was, in fact, one of several factors that led to the development of the Mobile Army Surgical Hospital. Experiments with Portable Surgical Hospitals showed the concept had promise.

The Mobile Army Surgical Hospital, or MASH as it quickly became known, was a new kind of organization that was announced on 23 August 1945 at the very end of World War II. The MASH was intended to bring emergency lifesaving surgery closer to critically wounded casualties. The concept called for placement of a sixty-bed, truck-borne MASH in support of each division in a forward location just out of enemy artillery range. The MASH was to be truly mobile, fully staffed with surgical and medical personnel, equipped to provide definitive life-saving surgery to make the patient transportable to rear medical facilities and to provide post-operative care for non-transportable patients. Five MASH units were created on paper between 1948 and early 1950, but were not staffed or ready for combat when North Korea invaded South Korea on 25 June 1950.

Three MASH units were established in Korea after hostilities began. They were staffed by personnel stripped from other Medical Department units, but not enough were in place to have one in support of each division as planned. The 400-bed Army evacuation hospital could not properly function in Korea because of the lack of transportation, an inadequate road and rail network and the volatile situation. In response MASH was enlarged to 150 beds in November 1950 and then to 200 beds in May 1951. The MASH concept of "surgery only" was abandoned under wartime pressure and the increase of wounded who needed treatment immediately. With this expansion in workload (medical cases in addition to surgery) and with an increase in personnel, rapid evacuation of patients to higher echelons was essential. In effect, the MASH became a small 200-bed capacity evacuation hospital that provided care to the division. In some instances a MASH exceeded 400 patients a day. Through December 1950 three MASH units supported four U.S. Infantry Divisions and other U.N. forces. By the end of 1950 there were four MASH units in support of seven divisions and attached foreign troops.

The United Nations Forces went on the offensive in 1951 and MASH units remained mobile and moved typically once per month. Through the latter of 1951 a concerted effort was made to move the MASH units closer to the battles, usually within 20 miles of the front line. This proved to be efficient for easy access to the wounded while still safely operational. Relatively inactive hospital staffs were moved to augment the heavily burdened MASH since it received the greatest casualty load. By 1951 there were five U.S. MASH units and a Norwegian Mobile Surgical Hospital (60 bed capacity) in support of U.S. and U.N. troops. An unstaffed MASH unit was held in reserve.

Standards for a MASH required that it be disassembled, loaded onto vehicles and ready to depart on six hour notice. It was operational within four hours after arrival at its new destination. Each MASH unit operated five surgical tables in a shift with a highly organized system to manage shock patients. An ambulance platoon was attached to each MASH to facilitate the rapid evacuation when post-operative recovery was complete. Additionally four helicopters were attached to each MASH. They, in turn, were utilized for resupply, rapid patient delivery to the MASH and comfortable evacuation from the MASH.

By 1952 fighting had stagnated and MASH units functioned primarily as static hospitals through July 1953 when a cease fire agreement ended the war.

The results were outstanding. The early treatment of wounded at a MASH located only minutes from the battlefield combined with swift, comfortable delivery and evacuation of the seriously wounded by helicopter, helped to lower the fatality rate for the Army's wounded. That rate had been 4.5 percent in World War II. In Korea it would eventually reach a new low of 2.5 percent.

The Mobile Army Surgical Hospital (MASH) concept was firmly established by its success in Korea. MASH units continued to serve and were deployed to Vietnam the 1991 Gulf War and the conflicts in Iraq and Afghanistan in the 2000's. At the same time, after a lag of 15-20 years, the Korean War MASH success in trauma management through enlisted medics (Para-Medics), helicopter evacuation and advanced methods in the treatment of shock became the model for civilian urban trauma centers.

As phenomenal as these MASH units were without the staff of surgeons, nurses, medical technicians and support personnel these MASH units would not have existed. Without so much as a Thank You the staff worked endless hours under extreme conditions, sometimes lacking the medical supplies needed to save the lives of military personnel. The MASH unit personnel did not get the recognition that they so rightly deserved and to call them anything less than a hero would be a travesty.

Engineer LeRoy Stelck

It is called The Forgotten War, because the Korean War came just five short years after the end of World War II. The service and sacrifice of those who served in the Korean War were quickly forgotten 15 months later when the Vietnam War broke out in November 1955. The Vietnam War has overshadowed it ever since.

Le Roy Stelck served his country as an engineer with the 712th Transportation Railway Operating Battalion. The soldier railroaders were some of the most unsung heroes of

The Forgotten War. Not only did they have to be capable to operate the railroad system in a foreign country, but they had to be able to protect the railway, keep it open and, if necessary, lay new track. In other words, they had to have the ability to fight and to make running repairs, all while transporting large quantities of troops, supplies and equipment to keep the UNC troops fighting. The Americans worked side-by-side with the Korean Railroaders.

In Korea the 712th was responsible for the operation of over 500 miles of railroad. They ran thousands of trains with inadequate and dilapidated equipment over lightweight and poorly maintained track, through crumbling tunnels and over bridges hastily constructed and repaired as the tides of battle moved back and forth. The 712th moved millions of tons of ammunition, equipment and food through the mountainous terrain and rice paddies just a short distance from the front lines and often under a threat of an enemy attack.

In April-May of 1953 Le Roy and the 712th took part in the Little Switch. At truce talks in Panmunjom both sides in the Korean Conflict agreed to an exchange of the sick and wounded prisoners of war. Trains were used in the exchange of prisoners. The Communist side returned 684 sick and wounded UNC soldiers, while the UNC returned nearly 6000 prisoners of war.

LeRoy's battalion also took part in the final exchange of prisoners, known as Big Switch, between August 5 and December 23, 1953. As reported by the American command: 75,823 Communist prisoners (70,183 North Koreans and 5640 Chinese) and 12,773 UNC prisoners (7862 South Koreans, 3597 Americans and 946 British) were returned.

LeRoy operated trains in both the Little Switch and the Big Switch. After the war LeRoy resumed his career with the Chicago Northwestern Railroad.

Jet Aircraft In Korea

Shaped in World War II by an increased concentration on the strategic role to attack an enemy's homeland, the Air Force now faced a conflict almost entirely tactical in character and limited as to how and where airpower could be applied.

The Far East Air Forces Fifth Air Force was the command and control organization for USAF forces engaged in combat. Tactical units conducted strikes on supply lines, attacked dams that irrigated North Korea's rice crops and flew missions in close support of United Nations ground forces.

Although President Truman did not want to risk extensive use of the U.S. bomber force which was used as a deterrent for possible Soviet aggression in Europe, a few groups of Strategic Air Command aging B-29 Superfortress bombers (that were not part of the nuclear strike force) were released for combat over the skies in Korea. Many of these B-29s were war-weary and were brought out of five years of storage. Although old and weary, they wreaked havoc on North Korean military installations, government centers and transportation networks.

As well as these B-29s performed, repeated bombing missions took their toll on the aging aircraft. The command looked for a source to replace the B-29s with an aircraft that could strike these installations in a close-in attack in a much faster manner. The jet age had entered the Korean War.

The Korean War was the last and only time, large numbers of piston-engine and jet-engine aircraft shared the war time skies. Korea marked the end of the line for major use of prop-driven combat aircraft of the active-duty USAF and brought in the jet age in real terms.

The first generation straight-winged F-80C Shooting Star and the F-84 Thunderjet jet aircraft were shown to be inadequate against the Soviet MiG-15s in actual dogfight situations. The F-80C was instrumental to gain and maintain air superiority over the Korean battlefield and rapidly cleared the skies of any North Korean aircraft that dared to venture into the air. However the straight-winged F-80s were inferior in performance to the MiG-15 and were soon replaced in the air superiority role by the swept-wing F-86 Sabre. Eventually the F-80s and F-84s were used as a strike force for close-in attack aircraft to perform attack missions that the B-29s had performed. However the swept-wing F-86 Sabre took control of the skies, which brought an entire new generation of swept-wing aircraft into the USAF arsenal in the 1950's.

The F-86 Sabre became the premier USAF fighter of the Korean War. By the end of hostilities, it had shot down 792 MiGs, with a loss of only 76 Sabre's -- a victory ratio of 10 to 1.

Women In The Korean War

After World War II, when the flag waving stopped and Johnny came marching home, G.I. Jane was out in left field without a ballgame. Millions of civilian women were literally kicked out of jobs and sent back to the kitchen. The war was over and there was no place for women in the military in the minds and hearts of many.

Fortunately a few visionaries had better sense than to let loose of all of the woman power that had rallied around the flag and served in the war. However in typical government fashion, politics prevailed and for three years the question of women as an integral part of the military disappeared.

One of the not so published issues was the fact that men did not want to ever have to take orders from women and heaven forbid if women became senior NCOs and Officers, heaven forbid this could happen.

General Eisenhower finally helped clear the way by strongly recommending that women become part of the U.S. military. He was backed by several other senior officers who had worked with women in WWII and could only praise their efforts.

On the 12th of June President Harry Truman signed on the dotted line and put Public Law 625 (The Woman's Armed Services Act of 1948) into effect. A law that today would be laughed out of town, it was so full of loopholes and strange parameters, but it opened the door for dedicated women to serve their country in peace time. One thing that it did not do that is often misinterpreted, is create separate women's branches, corps or forces. The only unit to retain that distinction was the WAC (Women's Air Corps). The rest of the women in the other branches of the service were, for all intents but not every purpose, fully integrated, or so the law implied. It just did not happen that way.

Two years later, in June 1950, as the overall number of women in the military dropped to a post-war low, the Korean War started and is now remembered as the Forgotten War. Over fifty thousand lives were lost in a country that we never heard of before, in a conflict termed a "limited war". President Truman ordered troops into South Korea and within a few days the Army Nurse Corps was also there. To the hundreds of women who served in Korea at the real Mobile Army Surgical Hospitals, it was no party.

When General MacArthur landed at Inchon, Army Nurse Corps officers also came ashore on the very same day of the invasion. The 13 Army nurses of the 1st MASH and those of the 4th Field Hospital made the landing and by the end of 1950 over two hundred Army Nurse Corps officers were in Korea.

In the Korean War Era over 120,000 women were on active duty. In addition to the nurses actually in Korea, many women served at support units nearby in Japan and other far eastern countries. Yet in research of women in war, it appears as though women who served in this campaign have become as forgotten as the war itself.

One of the women who served was Captain Lillian Kinela Keil, a member of the Air Force Nurse Corps and one of the most highly decorated women in the U.S. military. Captain Kinela flew over 200 air evacuation missions in WWII as well as 25 trans-Atlantic crossings. She went back to civilian flying with United Airlines after the war, but when the Korean conflict erupted she donned her uniform once more and flew several hundred more missions as a flight nurse in Korea.

As of the 1950's almost a million women had worn the uniform of the United States Armed Forces. They had been prisoners of war; had been wounded; they flew planes, planned strategies, nursed the casualties and died for this country.

The Vietnam War

The Vietnam War was the most unpopular war in our nation's history and the only war that our nation has ever lost. It has not been forgotten but those who served in it have. As unpopular as the Vietnam War was, like other wars, it produced hundreds, if not thousands, of heroes both in combat and outside of combat. The following unsung and unheralded heroes were just like any other forgotten or little known person from other wars. They were unknown for what they did and cast off as just someone who was involved in war but not in this author's eyes. They did their job, most times unnoticed, eventually they were forgotten but, to some, they are true heroes.

Vietnam War Photojournalists

In 1968 Catherine Leroy was one of the first female combat photojournalists of the Vietnam War Era. She surprised her North Vietnamese captors when she photographed and interviewed them as they returned her cameras as she was released her from detention. The photograph ended up on the cover of Life Magazine.

Many American soldiers, along with male correspondents, were shocked to see Leroy in 1966 when she landed in Vietnam on a one-way ticket from Paris through Laos to Saigon. She had her small Leica camera in hand. She was only 21 and her diminutive presence, at five feet tall and less than 90 pounds, didn't match the profile of the average foreign war correspondent.

Leroy entered a totally masculine world of war and war photojournalism to the point that it became a passion to her.

In Vietnam she was cool under fire and one of the few women photographers in the thick of the fighting and dying. Leroy wanted to show the war up close and personal. In one of her photographic sequences from Vietnam (1967) corpsman Vernon Wike applied first aid to a downed buddy, listened for a heartbeat and then looked up from the body with an anguished and confused look when he realized that the Marine was dead. The last photo shows the dead Marine alone with the landscape destroyed and the horizon blank. The photo series is a powerful reality check about the Vietnam War.

Leroy didn't just photograph the war from the sidelines -- she jumped in feet first, literally. When she joined up with the 173rd Airborne Division in 1967, thanks to a former boyfriend who taught her how to skydive. She jumped along with them in a combat operation to become the only known accredited journalist (male or female) to jump in combat with American troops at war.

Vietnam may have been the first conflict she covered with a camera, but she later covered many more including Afghanistan, Somalia, Iran and Iraq.

Sean Flynn

Sean Leslie Flynn was an American actor and freelance photojournalist who was best known for his coverage of the Vietnam War. Flynn, the only child of actor Errol Flynn and Lili Damita, disappeared in 1970 at age 28 while on assignment for Time Magazine and CBS News.

Flynn arrived in Vietnam in 1966 as a freelance photojournalist, first for a French magazine, then for Time Life and finally for United Press International. His photos were soon published around the world. He quickly made a name for himself as one of that group of high-risk photojournalists who would do anything to get the best pictures and even went into combat to do so.

In March of 1966, while in the field, Flynn was wounded in his knee. In mid-1966, he left Vietnam long enough to star in his last movie. He returned to Vietnam and made a parachute jump with the 1st Brigade, 101st Airborne Division in December 1966. In 1967 he went to Israel to cover the Arab-Israeli War. He returned to Vietnam in 1968 after the Tet Offensive, with plans to make a documentary about the War. In the spring of 1970 he went to Cambodia when news broke about North Vietnamese advances into that country.

On April 6, 1970 while traveling by motorcycle in Cambodia, Flynn and Dana Stone (again on assignment for Time Magazine and CBS News respectively) were captured by communist guerrillas at a roadblock on Highway One. They were never heard from again and their remains have never been found. Although it is known that they were captured by Vietnamese Communists Forces, it has been suggested that they died in the hands of "hostile" forces. Citing various government sources, the current consensus is that they were held captive for over a year before they were killed by Khmer Rouge in June 1971.

Dickey Chapelle

Dickey Chapelle was an American photojournalist known for her work as a war correspondent from World War II through the Vietnam War. Chapelle would go to extraordinary lengths to cover a story in any war zone. Chapelle was captured and jailed for over seven weeks during the Hungarian Revolution of 1956. She later learned to jump with paratroopers and usually travelled with the troops.

Despite early support for Fidel Castro, Dickey was an outspoken anti-communist and loudly expressed these views at the start of the Vietnam War. Her stories in the early 60's achieved great praise from the American Military Advisors, who were already fighting and dying in South Vietnam and the Sea Swallows, the anti-communist militia led by Father Nguyen Lac Hoa. Chapelle was killed in Vietnam on November 4, 1965 while on patrol with a Marine platoon in Operation Black Ferret, which was a search and destroy operation 16 miles or so south of Chu Lai, Quang Ngai Province, I Corps.

The lieutenant in front of her kicked a trip-wired booby trap that consisted of a mortar shell with a hand grenade attached to the top of it. Chapelle was hit in the neck by a two inch piece of shrapnel that severed her carotid artery and she died soon after.

Her body was repatriated with an honor guard that consisted of six Marines and she was given a full Marine burial. She became the first female war correspondent to be killed in Vietnam, as well as the first American female reporter to be killed in action.

Hugh Van Es

The end was at hand. Saigon swirled with panicked mobs of refugees desperate to escape before the North's invasion. On the outskirts of the surrounding city, more than a dozen North Vietnamese divisions prepared for their final assault on the outskirts of the surrounding city. A Dutch photographer, Hugh van Es, slipped through the crowds that day, snapped pictures and then hurried down Tu Do Street to the United Press International office to develop his film.

No sooner had he secured himself into the darkroom than a colleague, Bert Okuley, called out from the adjoining room, "Van Es, get out here! There's a chopper of the roof!" He pointed to an apartment building four blocks away where an Air America Huey, that was operated by the CIA, was perched. Twenty-five or so people scaled a makeshift ladder to try to climb onboard.

Van Es took ten frames from the balcony as the chopper over loaded with about 12 evacuees lifted off. Those left behind waited for hours for the helicopter to return, it never did. All day April 29, 1975 and into the evening, the sky was alive with choppers that darted in and out to at least four pick up sites in what was to be the largest helicopter evacuation in history. Van Es captured the last famous shots of the Vietnam War.

Van Es had taken dozens of memorable combat pictures during his seven years in Vietnam, but it was this one hurried shot from the balcony that brought him lifelong fame and became the defining image of the fall of Saigon and the end of the Vietnam War.

The Bird Dog

I have mentioned previously in this book of inanimate objects as being unsung or unheralded heroes. I felt it necessary to include the Cessna O-1 Bird Dog as one of these unheralded heroes because of its performance, not only in the Vietnam War, but also in previous wars.

The Cessna O-1 Bird Dog was not a glamorous, powerful, heavily armed aircraft such as the sleek fighter jets of the time, but was one of the most important aircraft to participate in any war.

It was developed as an experimental aircraft in 1947 and it soon proved to be an actuality. The U.S. Army awarded a contract to Cessna for 418 aircraft which was designated the L-19A Bird Dog. The prototype Cessna 305 first flew on December 14, 1949. Deliveries began in December 1950 and the aircraft was soon in use to fight in the Korean War from 1950 through 1953. As important as the Bird Dog was, it was an easy target. Its slow speed and the low altitude at which it flew made it an especially easy target for enemy small arms fire and surface to air missiles. It afforded little protection for the pilot when hit by small arms fire because of its light weigh (1600 pounds) and thin skin.

Very soon the Defense Department found out the importance of the Bird Dog and its use for the Korean War. Between 1950 and 1959 the Defense Department ordered 3200 L-19s. The aircraft were used in various utility roles, such as to spot for artillery and front line communications.

In 1962 the Army L-19 was re-designed the O-1 (observation) Bird Dog and entered its second war in Vietnam.

In the early 1960s the Bird Dog was flown by South Vietnamese airmen (ARVN-Army Republic Vietnam/ SVAF South Vietnamese Air Force), US Army aviators and clandestine aircrews. In 1964 the Department of Defense turned over its "fixed wing" O-1 Bird Dogs to the US Air Force, while the Army began its transition to "rotor-wing" force (helicopters).

In the skies over Vietnam, amid fast-moving attack aircraft that executed close air-support missions and helicopters that ferried troops in and out of remote jungle landing zones, an unobtrusive aircraft (0-1 Bird Dog) usually loitered in the skies to play a key, yet understated, role in the war.

The Bird Dog was used in Korea and in Vietnam. In both conflicts it was primarily employed as a platform for battlefield observation, to control artillery fire and close air-support direction.

The Bird Dog went through many transitions to improve its performance. It was originally fitted with a 210 horsepower engine. It was later changed to a 265 horsepower engine with an improved constant speed propeller.

With a maximum speed of 130 miles per hour, the Bird Dog could cruise 530 miles. The aircraft's on-station endurance and ability to loiter over a battlefield for up to five hours was most valuable to the infantry.

The Bird Dog could carry various payloads on four pylons under its wings. Usually the aircraft was outfitted with 2.75 inch white-phosphorus marking rockets to highlight the precise location where attack aircraft were to drop munitions or where helicopters could land.

For night missions the pylons could be outfitted with

two million candlepower flares to provide illumination for ground operations.

The O-1 Bird Dog pilots were the most decorated of any pilots in Vietnam and sustained the most casualties. At least two of them earned the Silver Star for their actions and nearly every Bird Dog pilot received at least one Distinguished Flying Cross.

In the course of the Vietnam War 469 O-1 Bird Dogs were lost to all causes. The USAF lost 178, the USMC lost seven and 284 were lost by the US Army, South Vietnamese Forces and clandestine operators. Three Bird Dogs were lost to enemy surface-to-air missiles (SAMs).

A Marine Corps General was quoted as saying, "The O-1 was the slowest aircraft in the Marine inventory, flew the lowest, drew the most fire and was by far the deadliest plane we used during the Vietnam War."

Gun Trucks: A Vietnam Innovation

Before 1967 there were no relevant doctrines for use of gun trucks to support convoys. Today our military forces in Iraq face an enemy that has chosen to attack soft targets, just as the enemy did in Vietnam. In Vietnam soft targets often were supply convoys that traveled with little protection. Our soldiers quickly began to use gun trucks to protect themselves and deter attacks. That is the situation our forces face today and they repeat many of the lessons learned in Vietnam.

In the fall of 1967 the North Vietnamese Army (NVA) decided to sever the lines of communication to the combat units at An Khe and Pleiku along route 19. Route 19, unknown to many U.S. drivers, had a fateful past. Thirteen years earlier in the French Indochina War, the Viet Minh completely destroyed an entire brigade-size French element along the same route. The U.S. forces were overly dependent on trucks for fuel and supplies, and the enemy commanders knew that.

September 2, 1967 a convoy of almost 40 vehicles from the 8th Transportation Group was in route from Pleiku. The convoy was split into two groups because of mechanical problems with a fuel tanker as it reached the An Khe Pass. It was almost dark when the lead gun jeep was ambushed. Simultaneously the rear half of the convoy was attacked and the fuel tanker began to burn. Many of the soldiers were unprepared and were caught by surprise. Enemy attacks on U.S. convoys had been minimal and limited to sniper-like attacks before the ambush. This was the first major ambush of an American convoy and it changed the nature of logistics operations for the rest of the war.

Initially machine guns were mounted on 2 ½-ton cargo trucks. Different configurations started to show up mainly due to commander-specific likes and ideas of what would work.

Some trucks were loaded with sandbags as protection but the rain storms, that were frequently experienced, soaked the sandbags and added additional weight to the trucks which made them too cumbersome. Drivers started to use steel plates to reinforce the vehicles.

Commanders decided to increase the number of gun trucks from 1 to 3 per 30- truck convoy. The improved gun trucks began to incorporate a second machine gun for added firepower.

It had been less than 90 days since the first large-scale ambush against U.S. convoys when the enemy made a second coordinated attempt. November 24, 1967 a convoy of almost 70 vehicles, that included six gun trucks and three gun jeeps, was attacked by the NVA. Along the 1000 meter convoy the enemy sustained 41 killed and four captured, while the convoy lost 14 trucks with two drivers killed. The additional gun trucks resulted in a severe blow to the enemy forces.

The truckers of the 8th Transportation Group are unsung heroes of the war. The gun truck, though not authorized on paper, was officially accepted and encouraged at all levels of the U.S. command in Vietnam.

Tunnel Rats

The Tunnel Rats were American, Australian and New Zealand soldiers who performed underground search and destroy missions in the Vietnam War.

They took the war to a one-on-one, eyes-to-eyes, I kill you or you kill me level.

In the course of the war the Viet Cong created extensive underground complexes throughout the country. The most famous complex was located in Chu Chi in Southern Vietnam.

Tunnel Rats were sent in to kill any hidden enemy soldiers and to plant explosives to destroy the tunnels whenever troops would uncover a tunnel. The Tunnel Rat was equipped with only a standard .45 caliber pistol, a bayonet or knife and a flashlight. Because of the confined space, the Tunnel Rats disliked the intense muzzle blast of the comparatively large .45 caliber round which would often leave them temporarily deaf. Most Tunnel Rats were allowed to choose another pistol in which to arm them.

The tunnels were very dangerous because of numerous booby traps and enemies that laid in wait. Often there were flooded u-bends in the tunnels to trap gas. Guards manned holes on the sides of tunnels through which spears could be thrust to impale a crawling intruder. Not only were there human enemies, but also dangerous creatures, such as snakes (including venomous ones), scorpions, spiders, ants and rats. Occasionally a tunnel rat would encounter a dead body that blocked the tunnel and hampered further forward progress. In this case the Tunnel Rat would have to back out of the tunnel, retrieve a length of rope to tie to the feet of the dead Viet Cong and pull the body from the tunnel.

The tunnels were such a big part in terms of how the Vietnamese fought. They served as escape routes and places to hide for the guerrilla-style Viet Cong fighters. They also were used to transport wounded, store and move weapons, munitions, food caches and other supplies and equipment.

The tunnel network was most extensive in the southern part of the country, which had hundreds of miles of underground passages and bunkers, some were large enough to serve as makeshift hospitals.

Tunnel Rats were generally, but not exclusively, men of smaller stature in order to fit in the narrow tunnels. Standard procedure required three men in the tunnel at a time. The amount of time spent in the tunnel depended on the complexity of the tunnel and they could be in and out in five minutes to as much as two hours. However long it was, to most it seemed like an eternity.

Tunnel Rats... the quietest, calmest, deadliest soldiers to serve in Vietnam.

The U.S. Coast Guard In Vietnam

The Coast Guard's involvement in the Vietnam War began in March of 1965 with the deployment of 26, 82-foot patrol boats in support of the U.S. Navy's Operation Market Time.

They served with quiet dedication and professionalism as the U.S. Coast Guard played a vital role to secure Vietnam's 1200 mile coastline. For the period of the Coast Guard involvement in Vietnam consisted of 8000 Coast Guard personnel and 56 different combat vessels that were assigned in country.

By the time the Coast Guard departed Vietnam, Coast Guard Cutters had logged over 5.5 million cruise miles, fired 6,000 gun missions and stopped and boarded 250,000 junks and sanpans. A truly outstanding record by anyone's standards.

Twelve U.S. Coast Guardsmen were awarded the Silver Star (our nation's third highest decoration for valor) for Conspicuous Gallantry in Action in the course of the war. Truly, U.S. Coast Guardsmen were the unsung heroes of the Vietnam War.

Women In Vietnam

All women who served in Vietnam were volunteers, either civilian or military. At one time there was discussion to draft women nurses, but this was never implemented. Some women military asked to go to Vietnam. Some were sent against their wishes, and even against the recruiter's promises, but all were volunteers because they voluntarily enlisted in the military.

They could not fight, nor were they allowed to carry weapons to defend themselves. Most were part of the pioneering Women's Army Corps (WAC), created in 1942 to integrate the Armed Forces. All of them enlisted for service in Vietnam, mostly in the early part of the war.

Women served in Vietnam in many support assignments: in hospitals, crewed on medical evacuation flights, with MASH units, on hospital ships, in operations groups, in information offices, in service clubs, headquarters offices and numerous other clerical, medical, intelligence and personnel positions.

There were women officers and enlisted women. There were youngsters in their early twenties with barely two years in service and career women with over forty years of service. In many cases women suffered the same hardships as the men and were often in the line of fire from rockets and mortars, particularly during the Tet offensive with the Viet Cong attacks on Saigon.

The U.S. Department of Veterans Affairs knows exactly how many men served in Vietnam (2,594,200) and how many were killed in action (58,188). It can furnish all kinds of stats about those soldiers, like the percentage of men who worked in supply (between 60 and 70 percent) as opposed to combat (30 to 40 percent). But ask about the women who served in Vietnam and the numbers disappear.

What is truly unconscionable in the annals of American military history is the fact that little or no data exists on the women who served, and yes, were injured or killed in Southeast Asia in the Vietnam War.

Accurate records are non-existent on how many women were there, what decorations they earned, where they served (and most important) what after-effects they have suffered and continue to suffer.

Some records were thrown together shortly after the war to number the amount of women in Vietnam but how accurate these records are is questionable:

- Over 500 WAC's were stationed in Vietnam.
- Over 200 Women Marines.
- Over 600 Women in the Air Force.
- Over 6000 Army, Navy, Air Force nurses and Medical Specialists.
- Untold numbers of Red Cross, Special Services, Civilian Service and countless other women were there.

Like a lot of Vietnam Veterans, these women have been dogged by their experiences in country. But unlike many veterans, they do not feel officially recognized and have been reluctant to seek help. Some have been plagued by symptoms of post-traumatic stress disorder and exposed to chemicals. While others have harbored the fact of their service like a shameful secret.

The USO Club Women

In 1974 the last USO Club in Vietnam closed forever and to this very day people still are amazed to learn that American women volunteered for civilian service in Vietnam with the USO Clubs.

The USO was created by President Franklin Roosevelt in 1941 to meet the morale needs of the service men and women in World War II. The USO was composed of six member agencies who gave financial and personal support to assist the role of the USO. These agencies were the YMCA, YWCA, Jewish Welfare

Board, Salvation Army, Travelers Aid and National Catholic Community Services.

Women from all walks of life signed up to be "VICTORY BELLES" to staff the clubs, run the canteens and sew buttons on uniforms. Vietnam had its share of Victory Belles, but they were called Associate Directors and Directors. The role was the same, bring a touch of home to our soldiers wherever they go and wherever they are. As the USO's logo says,"A Home Away From Home."

Vietnam may have been a different war with all front and no rear, but the danger was the same. Not only was it a messy war and increasingly controversial, it required the USO's civilian activities to become a vital arm of the national organization.

The first club, the Saigon USO, opened in April of 1963. It was to be an alliance that stretched eleven years to 1974 when all the women with the USO Clubs left Vietnam. It must be noted that the USO Clubs and the USO shows were two separate functions of the national USO.

As the war escalated, so did the need for USO Club staffers. Many young women with degrees in theatre, broadcasting and recreation were recruited directly from college campuses. They knew the danger and risk, but these young women signed up anyway and for the longest tours in Vietnam of 18 months.

They were not required to wear a uniform civilian clothes only and in this case, the mini skirt. No slacks were allowed and perfume was a must. Upon arrival in Vietnam they were told the men were never to see them cry. The director told them, "Their job in Vietnam was to be happy. Never let the men see you cry." The job was quite stressful at times, especially for the younger women who never faced this type of situation to deal with war weary soldiers. In order to keep the women happy and help them deal with the daily stress of the job they were given a week of R&R every three months.

Since the request for volunteers was granted but the Department of Defense, all orders were cut by said agency. Each volunteer was given the rank of GS-10 or captain. Her airfare was paid by the national USO. However, if the volunteer wanted out of her assignment earlier than the eighteen month tenure, she would have to pay her own way home.

At the peak of the war the USO had 22 clubs in Vietnam. As the war dwindled, the clubs did likewise to 17. Each club was self-supporting and offered snack bars, barbershops, gift shops, overseas telephone line, photo labs and hot showers. All clubs had three volunteers at all times for the duration of the war.

To cheer up military personnel in the remotest areas or on warships in the China Sea, the USO Club women brought programs to them; a Christmas party, a rock group or a Texas style barbecue. Since many civilian communities in Vietnam were off limits to uniformed personnel due to the danger of terrorist attacks, the USO Clubs offered the only safe and sane diversion. The familiar red, white and blue USO sign, with six stars to depict the six agencies, would mark the place where a fast-food snack, a milkshake, a chess game with a buddy or just to read hometown newspapers were there for everyone.

Through all this and more, the USO women stuck it out. Although most civilians at home came to know the USO only through televised replays of the Bob Hope Christmas shows, the real unsung heroes were the staff people who put everything they had into the Vietnam effort around the clock, month after month, year after year. The morale of the USO Club women never wavered.

Vietnamese Women At War

Are heroes only known by the country for which they serve? Of course they are. Can a person from a foreign land recognize a person or a group of people as heroes if they are not from that country? Why not?

Though the women from the United States served their country in many different scenarios women from other countries were not so fortunate.

For as long as the Vietnamese people fought against foreign enemies, women were a vital part of the struggle. The victory over the French at Dien Bien Phu is said to have involved hundreds of thousands of women, and many of the names in Viet Cong unit rosters were female. These women lived out the ancient saying of their country, "When war comes, even women have to fight."

Women from the countryside and Hanoi fought alongside their male counterparts in both the Viet Cong and North Vitenamese military in their wars against the South Vietnamese government and its French and American allies from 1945 to 1975.

These women were wives, mothers, daughters and sisters of men recruited into military service and because the war lasted so long, women from more than one generation of the same family often participated in the struggle. Some learned to fire weapons, lay traps, serve as village patrol guards and intelligence agents while others were propagandists and recruiters or helped keep the supply lines flowing.

Called the "Long Haired Warriors" they carried provisions through the jungles at Dien Bien Phu to the female Tunnel Rats at Chu Chi in the south. The Vietnamese women were the unsung heroes of the Vietnam's war of the national liberation.

Women POW's In Vietnam

There were five known women POW's in the Vietnam War. Accounts of others exist but are unconfirmed.

The first known female prisoner of war was Eleanor Ardel Vietti. She was captured at the leprosarium in Ban Me Thuot, May 30, 1962. She is still listed as a POW since confirmation of her death has never been established.

Monika Schwinn, a German nurse, was held captive for three and a half years and at one time the only woman prisoner at the "Hanoi Hilton."

The following missionaries were POWs:

Evelyn Anderson. Captured and later burned to death in Kengkok, Laos, 1972. Her remains were recovered and returned to the U.S.

Beatrice Kosin was captured and burned to death in Kengkok, Laos, 1972. Her remains were recovered and returned to the U.S.

Betty Ann Olsen was captured in a raid on the leprosarium in Ban Me Thuot during the Tet offensive in 1968. She died in 1968 and was buried somewhere along the Ho Chi Minh trail by fellow POW, Michael Binge.

The Medevacs Of Vietnam

The exploitation of the helicopter in the Vietnam War for medical evacuation was in full swing. Combat search and rescue helicopters retrieved downed pilots within hours of being shot down. Helicopters from the Marines and Army forces picked up the wounded soon after injury and quickly transported them to a treatment facility usually within minutes of being wounded. These Medevac helicopters were considered a significant factor in the decreased mortality from wounds in Vietnam. In WWII, about four percent of the casualties that reached medical treatment facilities died. This was reduced to two percent in the Korean War. The Vietnam War demonstrated fatality rates of one percent for casualties that arrived at medical treatment facilities.

South Vietnam is a country of mountains, thick triple canopy jungles and marshy plains, with few passable roads or serviceable railroads. Besides these natural and manmade obstacles, the Medevacs had to put up with torrential rains in the Monsoon Season which hampered rescue efforts. The allied forces waged a frontless war against a seldom seen enemy. Even more than in Korea, helicopter evacuation proved to be both valuable and dangerous. The physical problems of terrain and tropical climate were exacerbated by the Viet Cong's guerilla tactics which had no respect for the red crosses on the doors of the air ambulance helicopters.

In these conditions the modern techniques of air ambulances developed and matured. The physical obstacles of mountains, jungle and floodplain could be overcome only by helicopter. The frontless nature of the war also made the helicopter necessary for medical evacuation. Air ambulance units found ever wider employment as the helicopter (used both as a fighting machine and as a transport vehicle) came to dominate many phases of the Vietnam War.

A radio call by a field unit for a medical evacuation, or "Dustoff" as it was known, crammed the air waves all over the country every single day. Medevacs transported the sick, injured and wounded soldiers and Marines who required rapid movement to a medical facility. Also many Vietnamese civilian and military casualties were transported. Casualty proportions varied over the course of the war. Before 1965 about 90 percent of the patients were Vietnamese. After the US buildup began in 1965, the figure dropped to only 21 percent for 1966. As the United States started to turn over more of the fighting to the South Vietnamese in the 1960's, the Vietnamese number rose again until it reached 62 percent in 1970.

Although only about 15 percent of cases treated by Army medical personnel in the war were wounded in action, it seems that the percentage of wounded among the air ambulance patients was much higher, between 30 and 35 percent. The ambulances gave first priority to patients in immediate danger of loss of life or limb, which are conditions most closely associated with combat wounds. Up to 120,000 of the US Army wounded in-action that were admitted to some medical facility, 90 percent of the total were probably carried on the helicopters. This is about one third of the over 390,000 Army patients that the air ambulances carried to a medical facility.

Statistics confirmed that air ambulance pilots and crewmen were at risk to be injured, wounded, or killed in their one year tour. About 1400 Army commissioned and warrant officers served as air ambulance pilots in the war, one of the most dangerous types of aviation in Vietnam. About 40 aviators (both commanders and pilots) were killed by hostile fire or crashes induced by hostile fire, 180 were injured or wounded as a result of hostile fire, another 48 were killed and about 200 injured as a result of non-hostile crashes of which many happened at night and in bad weather. This was in spite of the fact that they were warned that air ambulance work was a good way to get killed.

In 1969 the Defense Department started a program with state agencies to provide emergency medical personnel and medevac helicopters to respond to serious traffic accidents in the states. The program was called MAST (Military Assistance to Safety and Traffic). Military helicopters assigned to the MAST program were painted white with large red crosses on the sides.

The Medical Services Command in Vietnam got the idea that medevac helicopters were shot at in Vietnam because they were olive drab and looked too much like other choppers that were strafing enemy positions and units all over the country.

The order came down from command that some medevac helicopters should be painted white like the stateside MAST helicopters and maybe the enemy would stop shooting at them! Needless to say medevac units chosen to receive these white Hueys were overjoyed with this decision.

The 57th Med Detachment began to take delivery of these white Hueys in January 1972. The flight crews quickly dubbed these aircraft "White Elephants." The nose retained the flat black paint, with the lower half of the nose painted white. The skids were olive drab. The 57th Med Hueys had white skid tips. As expected, the NVA and the VC targeted these white helicopters as quickly as other choppers.

Medevac Pilots

I don't usually point out an individual person for their heroic actions but these three individuals flew Huey Medevac helicopters in Vietnam and their heroics deserve to be mentioned.

Michael J. Novosel

Michael J. Novosel had a long illustrious military career that spanned over 44 years. In 1963 Novosel worked as a commercial airline pilot when patriotism called him to active military duty. He was 41 and the Air Force did not have space for any more officers in the upper ranks. It was then that Novosel made the decision to give up his rank of Lieutenant Colonel in the Air Force and joined the Army to fly helicopters as a Chief Warrant Officer with the elite Special Forces Aviation Section. He served his first tour in Vietnam and flew medevac helicopters (Dust-off) with the 283rd Medical Detachment. His second tour in Vietnam was with the 82nd Medical Detachment. On this tour Novosel flew 2543 missions and extracted 5589 wounded personnel, among them his own son, Michael J. Novosel, Jr. The next week Michael J. Novosel Jr. returned the favor and extracted his father after he was shot down.

On the morning of October 2, 1969 he set out to evacuate a group of South Vietnamese soldiers who were surrounded by the enemy near the Cambodian border. The soldier's radio communication was lost and their ammunition was expended. Novosel flew in at low altitude without air cover or fire support while under continuous enemy. He skimmed the ground with his helicopter while his medic and crew chief yanked the wounded men on board. He completed 15 hazardous extractions, was wounded in a barrage of enemy fire and momentarily lost control of his helicopter that day, but when it was over he had rescued 29 men. Novosel completed his tour in March 1970. In 1971 President Richard Nixon placed the nation's highest award for valor in combat, the Medal of Honor, around Novosel's neck. He also received the Distinguished Service Cross, the Distinguished Service Medal, Distinguished Flying Cross with two Oak Leaf Clusters, Bronze Star with Oak Leaf Cluster and the Purple Heart.

Bruce P. Crandall

Crandall was assigned to A Co., 229th Assault Helicopter Battalion. On November 14, 1965 he led the first major division operation of the Vietnam War as they landed elements of the 1st Battalion and 2nd Battalion of the 7th Cavalry Regiment and the 5th Cavalry Regiment of the U.S. Army into landing Zone X-Ray in the Battle of La Drang.

Crandall and Major Ed Freeman were credited with the evacuation of seventy wounded soldiers in the fierce battle of La Drang. Twelve of these fourteen flights were made after the Medevac unit refused to land in the intensely hot landing zone. Crandall's helicopters evacuated more than 75 casualties in a flight day that started at 6:00 a.m. and ended at 10:30 p.m.

The two flew in ammunition needed for the 7th Cavalry to survive. The aircraft they flew were unarmed.

Crandall, for his heroic feat, received the Distinguished Flying Cross which was upgraded to the Medal of Honor in February 2007.

A little known fact about Bruce Crandall is that he was portrayed by Greg Kinnear as the helicopter pilot who made one flight after another to evacuate the wounded in the 2002 movie We Were Soldiers that was the actual event that led to his Medal of Honor award.

Ed "Too Tall" Freeman

As a child Ed "Too Tall" Freeman dreamed of becoming a soldier when he grew up. Little did he know that his dream would become a reality and he would become one of the nation's few aviators to receive the Medal of Honor in the Vietnam War.

Freeman's military career started in World War II and eventually he served in the Korean War where he fought as an infantry soldier in the battle of Pork Chop Hill. He earned a battlefield commission to second lieutenant as one of only 14 survivors out of 257 men who made it through the opening stages of the battle.

The commission made him eligible to become a pilot which also was a childhood dream of his. However when he applied for pilot training, standing at 6′ 4″, he was too tall for pilot duty. The phrase stuck and he was known by the nickname of "Too Tall" for the rest of his career.

On November 14, 1965 Freeman and his unit transported a battalion of American soldiers to the La Drang Valley. Later when they returned to their base, they learned that the soldiers had come under intense fire and had taken heavy casualties. Enemy fire around the landing zone was so heavy that the medical evacuation helicopters refused to fly into the landing zone. Freeman and his commander, Major Bruce Crandall, volunteered to fly their unarmed, lightly armored UH-1 Huey in support of the embattled troops. Freeman made a total of fourteen trips under heavy enemy fire to the battlefield to bring in water and ammunition and take out wounded soldiers. This was later named the Battle of La Drang. By the time they landed their heavily damaged Huey, Captain Freeman had been wounded four times by ground fire. In 2001 President Bush presented Freeman with the Medal of Honor. Freeman was Bruce Crandall's wingman on the same day that Crandall earned his Medal of Honor.

Hugh Thompson Jr.

Everyone remembers the My Lai Massacre of 1968 which involved Second Lieutenant William Calley, Commander of 1st Platoon, of C Co., 1st Battalion and 20th Infantry. Calley was the one who had to answer to the massacre of human beings although there were many others involved in the actual killings of civilians at My Lai. What led to Calley's conviction was the testimony of a military officer who witnessed the killings, intervened and stopped further killings on that day.

Hugh Thompson was a helicopter pilot who flew an OH-23 Raven observation and reconnaissance helicopter.

On March 16, 1968 he and his crew supported Task Force Barker in a reconnaissance capacity. Specialist Glenn Andreotta served as door-gunner and crew chief. the other door-gunner was Specialist Lawrence Colburn.

In the early morning of March 16, 1968 Thompson's OH-23 did not encounter enemy fire over My Lai 4. He spotted two possible Viet Cong suspects, forced the Vietnamese men to surrender and flew them off for a tactical interrogation. Thompson also marked the location of several wounded Vietnamese with green smoke, a signal that they needed help.

When he returned to the My Lai area at around 0:900 after refueling he noticed that the people he had marked were now dead. He marked the location 200 meters south of the village of a wounded Vietnamese woman out in a paddy field beside a dike. Thompson and his crew watched from a low hover as captain Ernest Medina (Commanding Officer of C Company) came up to the woman, prodded her with his foot and then shot and killed her.

Thompson then flew over an irrigation ditch that was filled with dozens of bodies. He was shocked at the sight and he radioed his accompanying gunships. He knew his transmission would be monitored by many on the radio net. "It looks to me like there's an awful lot of unnecessary killing going on down there. Something ain't right about this. There are bodies everywhere. There's a ditch full of bodies that we saw. There's something wrong here."

Movement from the ditch indicated to Thompson that there were still people alive in there. Thompson landed his helicopter and dismounted. David Mitchell, a sergeant and squad leader in 1st platoon, walked over to him. When asked by Thompson whether any help could be provided to the people in the ditch, the sergeant replied that the only way to help them was to put them out of their misery. Calley then came up and the two had the following conversation:

Thompson: "What's going on here lieutenant?"
Calley: "This is my business."
Thompson: "What is this? Who are these people?"
Calley: "Just following orders."
Thompson: "Orders? Whose orders? "
Calley: "Just following..."
Thompson: "But, these are human beings, unarmed civilians, sir."
Calley: "Look Thompson, this is my show. I'm in charge here. It ain't your concern."
Thompson: "Great job."
Calley: "You better get back in your chopper and mind your own business."
Thompson: "You ain't heard the last of this!"

Thompson took off again and Andreotta reported that Mitchell (C Company Commander) was now executing the people in the ditch. Furious, Thompson flew over the northeast corner of the village and spotted a group of about ten civilians including children, who were running toward a homemade bomb shelter. Soldiers from the 2nd Platoon, C Company were in pursuit of them. Thompson realized that the soldiers intended to murder the Vietnamese and he landed his aircraft between the soldiers and the villagers. Thompson turned to Colburn and Andreotta and told them that if the Americans began to shoot at the villagers or him, they should fire their M60 machine guns at the Americans. "Y'all cover me! If these bastards open up on me or these people, you open fire on them. Promise me!" He then dismounted to confront the 2nd Platoon's leader, Stephen Brooks. Thompson told him he wanted help to get the peasants out of the bunker.

Thompson: "Hey listen, hold your fire. I'm going to try to get these people out of this bunker. Just hold your men here."

Brooks: "Yeah, we can help you get 'em out of that bunker-with a grenade!"

Thompson: "Just hold your men here. I think I can do better than that."

Brooks declined to argue with him, even though as a commissioned office he outranked Thompson.

After he coaxed the 11 Vietnamese out of the bunker, Thompson persuaded the pilots of the two UH-1 Huey gunships that flew as escort to evacuate them. As Thompson was on his way to the base to refuel, Andreotta spotted movement in an irrigation ditch that was filled with approximately 100 bodies. The helicopter landed again and the men dismounted to search for survivors. After they waded through the remains of the dead and dying men, women and children Andreotta extracted a live boy named Do Ba. Thompson flew the survivor to the ARVN Hospital in Quang Ngai.

On return to their base at about 1100 Thompson heatedly reported the massacre to his superiors. His allegations of civilian killings quickly reached Lieutenant Colonel Frank Barker, the operation's overall commander. Barker radioed his executive officer to find out from captain Medina what was happening on the ground. Medina than gave the cease-fire order to Charlie Company to "knock off the killing".

Thompson made an official report of the killings to command. Concerned senior officers of the Americal cancelled similar planned operations by Task Force Barker against other villages (My Lai 5, My Lai 1, etc.) in Quang Ngai Province. This possibly prevented the additional massacre of hundreds, if not thousands, of Vietnamese civilians.

Initially commanders throughout the America's chain of command were successful to cover up the MY Lai Massacre. Somewhat perversely, Thompson quickly received the Distinguished Flying Cross for his actions at My Lai. The citation for the award fabricated events and praised Thompson for taking a Vietnamese child to a hospital caught in intense crossfire and said that his sound judgment had greatly enhanced Vietnamese-American relations in an operational area. Thompson threw the citation away.

Thompson continued to fly observation missions and was hit by enemy fire a total of eight times. In four of those instances his aircraft was lost. In the last incident his helicopter was brought down by enemy machine gun fire and he broke his back in the crash landing. This ended his combat career in Vietnam and he was evacuated to a hospital in Japan to begin a long period of rehabilitation. He carried the psychological scars from Vietnam for the rest of his life.

When news of the massacre publically broke, Thompson repeated his account to Colonel William Wilson and Lieutenant General William Peers in their official Pentagon investigations. In late 1969 Thompson was summoned to Washington DC and appeared before a special closed hearing of the House Armed Services Committee. There he was sharply criticized by Congressmen, in particular Chairman Mendel Rivers (Democrat S.C.), who were anxious to play down allegations of a massacre by American troops. Rivers publically stated that he felt Thompson was the only soldier at My Lai who should be punished for turning his weapon on fellow American troops, but was unsuccessful in his attempt to have him court-martialed. As word of his actions became publically known Thompson started to receive hate mail, death threats and mutilated animals on his doorstep.

Acts of courage do not necessarily make a hero. In the case of Hugh Thompson he did what he thought was the right thing to do and never thought of the long term consequences or the danger he faced at the scene when he faced his own troops. Hugh Thompson did what he thought was right and he turned out to be a hero, but not in everyone's eyes. Those who have weighed all of the facts certainly believe that Hugh Thompson is truly an American hero. Those who do not believe, I am sorry for them.

The Secret Hidden War

The following may be quite shocking to many people who were not involved in the Vietnam War, but for those who were there it may come as no surprise.

Laos was officially neutral when the Vietnam War broke out. The U.S. had signed an international agreement (the Geneva Accords) which intended to keep Laos neutral and prevent fighting there.

In reality this agreement gave the Communists the upper hand, as they flagrantly violated the agreement. In response to the presence of active North Vietnamese troops in Laos, the U.S. tried to oppose them without appearing to violate the Geneva Accords. They secretly recruited freedom-loving locals to fight the Communists and these freedom-loving locals were Hmong.

Most Americans, and a good number of them fighting in Vietnam, thought that Laos was not part of the Vietnam War. But Laos played a critical role, especially since supplies from North Vietnam to its warring troops primarily moved along the Ho Chi Minh trail that passes through Laos.

Much fighting occurred along this trail and the surrounding regions in Laos. But our military efforts there were not publicized to avoid international criticism. So we pretended that nothing was happening in Laos, while North Vietnamese troops were actively helping the Pathet Lao (Lao Communists) take over the country. Thousands of poorly equipped Hmong were fighting a war against terrible odds. Many Hmong lives would be lost in the unpublicized battles of Laos.

In 1963 the Kennedy Administration had the CIA increase the secret Hmong Army in Laos to 20,000 soldiers. Significant battles occurred as the North Vietnamese and the Pathe Lao occupied major areas in northern Laos in 1964.

Meanwhile the U.S. began a secret air war in Laos. By 1968 US pilots would fly 3000 dangerous sorties a day to battle many thousands of Communist troops. Hmong soldiers rescued many American pilots who were shot down. Sometimes dozens of Hmong would die in order to rescue one American pilot. As the Vietnam War escalated so did the secret war in Laos.

For nearly 40 years the secret war in Laos was just that, SECRET. The Vietnam War (Laos) missions were so secret that the top U.S. brass could not, and would not, talk about them until about 2003.

Thousands of American soldiers fought in rice paddies and the jungles of Vietnam, but the men of the Studies and Observation Group (SOG) fought a different war against the North Vietnamese Army.

SOG recon teams generally consisted of three American Special Forces soldiers and six to ten allied Montagnard, Chinese Nung, Cambodian or Vietnamese soldiers. Recon teams struck 15 to 50 miles deep into Laos, Cambodia and North Vietnam. Few of them were officers. Most were enlisted men known as Green Beret volunteers.

They were given a briefing and told, "Now, you're all volunteers" and they were now on a classified secret mission into Laos.

About 2000 Americans and countless allies served in SOG between 1964 and 1972. More than 300 Americans were killed and an accurate account of allied soldier's loss does not exist.

A secret mission in 1970 to rescue allied and American POWs from the Son Tay prison, that was located 23 miles west of Hanoi failed. The North Vietnamese moved their prisoners around from time to time. On this particular day they had been moved a few days before the raid because of a potential risk of flooding by the nearby river. The American command knew the North Vietnamese had probably moved the prisoners but the raid was still performed. Only one American soldier was wounded in the leg and one had broken his leg in an intentional aircraft crash in the middle of the prison compound. One hundred North Vietnamese guards were killed and the prison was secured and quickly abandon. It was a disappointment not to find anyone there for the personnel that participated in the raid.

To discover what the enemy was doing SOG teams were armed to the teeth and moved quietly through the fields and jungles, while they whispered and evaded. If discovered, there was usually a quick and deadly firefight. If they were lucky they could escape by helicopter that was protected by American Fighter planes.

On one mission to rescue a downed pilot, they found him and as he was loaded onto a helicopter he replied, "I didn't know that we had American aviation here." He was promptly told, "We don't and you did not see us and forget what you have seen here".

Hundreds upon hundreds of missions were performed and thousands of aircraft sorties were flown into Laos between 1964 and 1972 by a select group of SOG soldiers and military pilots. I would say that this is probably one of the greatest groups of unheralded heroes of the Vietnam War.

The men of SOG were honored for their gallantry with the Presidential Unit Citation. More than 2000 were given medals for their heroism. Eleven of them received the Medal of Honor.

Honored But Not Forgotten

The Vietnam Veterans Memorial contains the names of 58,260 men and women who died while serving in the U.S. Armed Forces in the Vietnam War.

The Department of Defense developed specific parameters that allow only the names of service members who died of wounds suffered in combat zones to be added to the Vietnam Veterans Memorial. Under the Department of Defense "In Memory" program those men and women who have died prematurely as a result of the Vietnam War, but who do not meet the criteria, may be added. Many of their deaths are the result of Agent Orange exposure and emotional wounds that never healed.

In a ceremony on April 20, 2009 family members read aloud their loved one's names in chronological order by date of death. After the ceremony participants laid tributes at the base of The Wall next to the honoree's date of service in Vietnam. These Vietnam veterans came to rest near the comrades with whom they served. With 2009's honorees, over 1800 individuals will be honored in the In Memory Honor Roll.

"In Memory Day" it allows The Wall to do what it does best to provide a healing environment for family members and friends. It also allows America to pay tribute to these brave Americans who served and sacrificed for their country.

The Persian Gulf Wars

The Persian Gulf Wars and the unheralded heroes are a difficult subject to write about. These wars, above all, were wars of advanced technology that dictated who would and would not be killed by the new sophisticated machinery of war. Radar guided Smart Bombs dropped by our aircraft selected and singled out only enemy targets that were of significant military importance.

Operation Desert Storm (the first Persian Gulf War) though it lasted only 42 days, set a precedent of how a war should be handled and executed though those wars that followed did not follow Desert Storm's example. Saddam Hussein's army, though somewhat organized and equipped, could not contend with the Coalition's mite and advanced technology proved to be an easy adversary.

Earlier in this book I mentioned how not all heroes are people or animals and that an inanimate object may be classified as a hero if it shows that it advances the military might or helps the war in an extraordinary way.

Precision Guided Munitions

In war bombs and missiles played an important part of how the war progressed. Though a strategic bombing mission may have destroyed certain sought-after enemy targets, it also killed many innocent civilians along with the enemy. This was an unfortunate thing and it was not somewhat controlled until the early Gulf War (Desert Storm) where precision guided munitions were used.

Precision Guided Munitions (also known as Laser Guided Bombs) although they were missiles also used these laser guiding systems that directed the munitions to the correct target and did their job to destroy enemy targets without loss of civilian life. Statistics showed that these targets were usually within striking distance of some sort of civilian population and therefore innocent civilians became part of the war statistics. As journalists in their hotel rooms watched specific buildings in downtown Baghdad that could be bombed when precision guided cruise missiles flew to take out their intended target.

Though precision guided munitions helped immensely to limit civilian casualties, they only amounted to about 7.4 percent of all bombs dropped by the coalition. Other bombs dropped included Cluster Bombs, which dispersed numerous sub munitions that covered a very large area and Daisy Cutters, a 15,000 pound bomb that disintegrated everything within several hundred yards.

Though the last two munitions did the job as required, the unheralded hero in this bunch is the Precision Guided Munitions which did its intended job to take out the enemy position but limited civilian casualties.

Satellite Directed Systems

Global Positioning Systems (GPS) of today help us navigate the roads that we drive every day. Most do not know that the GPS, as we know it, was a very important system that helped the U.S. Military and Coalition Forces navigate easily across the desert. Communication through these devises helped to plan attacks at strategic locations that without them would not have been possible.

Airborne Warning and Control System (AWACS)

Airborne Warning and Control System, or AWACS as it is known, and satellite communication systems were a vital part of the technical communication link between ground forces, Air Forces and the Navy. Both were used in command and control areas of operations. It is one of the many reasons why the air war was dominated by the Coalition Forces.

Photocopiers

American-made color photocopiers were used to produce some of Iraq's battle plans. Some of the copies contained concealed high-tech transmitters that revealed their positions to America electronic warfare aircrafts which lead to more precise bombings.

The Patriot Missile

The U.S. Patriot Missile was used for the first time in combat in The Gulf Wars. The U.S. claimed a high effectiveness against Scud Missiles at the time, but later analysis gave figures as low as nine percent, with forty-five percent of the Patriot launches used against debris or false targets. Both the U.S. Army and the missile manufacturers maintained the Patriot delivered a "miracle performance" in the Gulf War to save many lives.

The Gun Truck: A Vietnam Innovation Returns

There were lots of changes to convoy doctrine in Vietnam. Experimentation with dividing convoys into ten trucks with a gun truck was tried. Because of the Viet Cong's ambush on November 24, 1967 a change had to be made and the command decided upon an optimum of 30 trucks in a convoy with an adequate number of gun trucks as support.

There were lots of experimentation with gun truck designs:

quad 50 caliber machine guns, APC (armed personnel carriers) trucks and V-100 armored cars. The 5-ton with a gun box proved to be the best. As late as 1969 senior leaders thought the V-100 would replace the gun truck. The drivers did not like the idea of being crammed into a confined space. The armored car had limited visibility and fire power. If an enemy round penetrated the armor and entered the vehicle it would ricochet around inside. As the gun truck design improved, the crews gained greater confidence in them. Then they began to drive into kill zones to protect disabled vehicles and rescue drivers.

Unfortunately the lessons learned in Vietnam were forgotten during the Cold War. It was not until 1993, in Somalia, that the Army once again began to armor support vehicles. After Somalia, however, the wake-up call in Iraq came at An Nasariyah in March 2003 with the heavily reported ambush of the 507th Maintenance Company. In the following months the Iraqi insurgency stepped up attacks against the relatively unprotected convoys that left Kuwait. The result was that many students of Vietnam gun trucks and doctrine began to turn to the lessons of the past. Because of those lessons learned on battlefields almost 40 years ago, the military was able to quickly adapt Vietnam-era doctrine and integrate new techniques to combat the Iraqi insurgency. The result is that logistics convoys are no longer an easy target and that, if they are engaged, convoy escort platforms (as they are now known) can inflict heavy damage on an enemy force.

Many lives have been saved, both in Vietnam and in Iraq, because of the gun truck doctrine developed in Vietnam. Convoys cannot be expected to travel as freely as they once did through the battlefields of World War II, Korea, or even Desert Storm. The enemy knows soft targets wherever it can find them. Thanks to the gun truck, the targets will no longer be our convoys.

Many technological advanced weapons were used in the war and in the subsequent Gulf Wars: aircraft, fixed wing and helicopters, ground attack vehicles such as tanks, Bradley fighting vehicles, advanced personnel carriers and of course artillery and small arms.

The advancement in small arms in the Gulf Wars was beyond belief. Each soldier not only had a fully automatic weapon (rifle) but this weapon also had a grenade launcher attached. The advancement of the machine gun known as a crew served weapon (a weapon manned by more than one person) was now a weapon that was controlled by one man.

A sniper rifle was developed that could take out an adversary at more than one mile without compromising the shooter's position.

Enough of the inanimate heroes of war, let's get to those who used these advancements of war.

Women In The Gulf Wars

Mobilization for the Gulf War included an unprecedented proportion of women from active forces (7 percent) as well as Reserve and National Guard (17 percent). It was the largest deployment in U.S. history.

Over 40,000 were deployed and several thousand more served stateside in essential mission support roles. The service women of the 90's served in the mainstream of the mission goals of Desert Storm and demonstrated that women perform as well as men. The so-called art of war became so much more technological and with so much less individual ground combat, the exclusion of women from any position in the military is ludicrous. The old-fashioned rationale of military planners and congressional leaders has to catch up with the advent of the 21st century.

Women in Desert Storm did everything the male troops did except engage in ground combat. They could essentially get fired upon. They just weren't theoretically allowed to shoot back by existing regulations.

However here is an excellent quote on the way it really was:

"I was a female paratrooper with the 82nd Airborne Division during Desert Storm/Desert Shield. I want to make you aware of the fact that females in the 82nd were among the ground troops that pushed into Iraq during the ground war.... And we most definitely could shoot back."

More than 40,000 women were deployed for the Desert Storm/ Desert Shield. Fifteen were killed and two were imprisoned by Iraqi forces. Media coverage revealed how completely U.S. service women integrated into almost every military unit.

Women drove and repaired trucks, dug and built bunkers, flew jet tankers, refueled bombers and fighters in midair, launched Patriot missiles and endured enemy attack. They guarded POWs, lived with, worked with and commanded male soldiers.

On the evening news Americans saw military women in the Persian Gulf War who transported troops and supplies. A woman led a company of Chinook helicopters into Iraq on the first day of the ground war. Women served aboard Navy hospital ships and destroyer tenders. They commanded units from graves registration detachments to battalion-sized material management centers.

Women proved they were able to work beside men and get the job done. The operations established new frontiers for women in combat even though federal law prohibited women to serve in direct combat and barred Navy and Air force women from combat ships and aircraft. When troops of Desert Storm and Desert Shield began to come home, there were celebrations, fanfare, yellow ribbons and an appreciative flag-waving public. Congress began to rescind the statutory restrictions which had banned servicewomen from combat aircraft and vessels.

Currently servicewomen serve aboard every kind of military aircraft and naval vessel with the exception of submarines. No law bans women from armed combat, although the Army still maintains policies which limit women's combat exposure.

The following women cannot be classified as unsung or unheralded heroes but because of their bravery under fire, they are in a class of true American heroes, women combat soldiers.

Leigh Ann Hester

Hester was the first female soldier to receive the Silver Star award for exceptional valor since World War II and the first ever to be cited for valor in close quarters combat.

Hester's squad of two women and eight men in three Humvees was shadowing a 30-truck supply convoy when approximately fifty insurgent fighters ambushed the convoy with AK-47 assault rifles, RPK machine gun fire and rocket propelled grenades (RPGs).

Hester maneuvered her team through the kill zone and into a trench line with hand grenades and grenades launcher rounds. Hester and her squad leader, Timothy Nein, assaulted and cleared two trenches. In the 25 minute firefight Hester killed at least three enemy combatants with her M4 rifle.

When the battle was over 27 insurgents were dead, six were wounded and one captured.

Monica Lin Brown

Monica Lin Brown, a combat medic, became the first women in Afghanistan since World War II to receive the Silver Star, the United States' third highest medal for valor.

After a roadside bomb detonated near a convoy of Humvees in the eastern Paktia Province of Afghanistan, Brown saved the lives of fellow soldiers as she used her body to shield wounded soldiers by running through insurgent gunfire while mortar rounds fell nearby. Because females are not allowed in direct action operations, Brown was pulled back to the base at Khost shortly after the incident.

She was presented with the Silver Star by Vice President Dick Cheney in March of 2008.

Lori Hill

Chief Warrant Officer 3 Lori Hill, a pilot, was in her Kiowa Warrior helicopter when the lead chopper came under heavy fire. She drew the fire away and simultaneously provided suppressive fire for the troops engaged with the enemy on the ground.

A rocket propelled grenade (RPG) hit her and damaged the helicopter's instrumentation, but instead of focusing on her predicament, she established communication with the ground forces and continued to provide them with aerial weapon support until the soldiers reached safety.

As she turned her attention to the aircraft, which was losing hydraulic power, the aircraft took on machine-gun fire and a round crashed into one of Hill's ankles. Still with a damaged aircraft and an injury, she landed her helicopter safely at Forward Operating Base Normandy and saved the lives of her crew and aircraft.

For her actions she was awarded the Distinguished Flying Cross by Vice President Dick Cheney.

Many women performed acts of bravery such as these in many different capacities during the Gulf Wars.

Women in combat? There have always been women in combat. Gender does not make a person a hero, no more than the color of their skin. It is what lies within a person's heart and character that makes a hero.

Women POW's Of The Gulf Wars

Operation Desert Storm saw the capture and imprisonment of an Army Flight Surgeon, Major Rhonda Cornum and an Army Transportation Specialist SP4, Melissa Rathbun-Nealy.

Operation Iraqi Freedom had two women prisoners of war. Supply clerk PFC Jessica Lynch, 19, was taken prisoner after her unit was ambushed. Ten days later, she was freed in a daring raid by U.S. Marines.

Army Specialist Shoshana Johnson, 30, was a prisoner in Iraq before she and six of her fellow soldiers were rescued on April 13, 2003.

To be sure there are many more women who have been prisoners of war including military and civilian women from nations around the world from wars long forgotten and from covert operations never revealed. They have been denied recognition, denied awards and decorations and denied their rightful place in history. The American military refuses to acknowledge their combat status. The American public thinks it never happened.

Journalists In War

There will most assuredly be war raging somewhere in the world at any one time and where there is war there will be those who report about it. Journalists and the media cover these wars from the start until the end so the public can be aware of what is going on in the war every day. They face the same danger of those who are waging the war.

The War on Terror, Desert Storm and Desert Shield continues to claim the lives of soldiers, innocent civilians and journalists. Experts say that war reporters have a lot to learn on how to protect themselve as they try to get their story.

Too many times journalists are the only professionals on the battlefield, or in a disaster zone, and are quite unprepared for what they will encounter.

Since the start of the War on Terror hundreds of journalists have gone there to try to cover the war. In 2010 alone 46 journalists were killed in hostile environments to report events that include Iraq and Afghanistan, compared to the Vietnam War, which claimed about 70 journalists.

Technological innovation and smaller, lighter equipment has made war reporting more dangerous than ever before. Now more and more reporters cover the news from the front lines, including camera operators.

To live up to the networks standards of " the story must be captured", this trust between the reporters and the military must be necessary. But consequently, a reporter's safety is often at risk. However despite how dangerous war reporting can be, it is an essential job that someone must do. Their job is to keep the outside world informed regardless of the danger.

Close to 230 media staff, of whom 172 were journalists, died in Iraq as of a result of the conflict stemming from the American intervention between March 20, 2003 to August 19, 2010, when the US Army's last combat brigades withdrew from Iraq.

Not all media staff killed were from the US. A good number of them were from the coalition. Ninety-three percent were men and seven percent were women.

War reporters are the unsung, unheralded heroes behind most of the news footage we see on our screens every day.

Someone once said, "There is no guarantee in War.... They must ask themselves, is this story worth the risk? Maybe, but no story is worth getting killed for."

The True Unheralded Heroes

Throughout this entire book I have talked about unsung, unheralded and famous heroes who performed acts of heroism that earned them great awards. What I have not done is made reference to, or talked about, probably the most forgotten and less thought about heroes of all: the husbands, wives and families of those who fight for our right to be free, the American Soldier.

There is talk about what our fighting soldiers endure on a daily basis as they fight the War on Terror, but no one talks or thinks about those who are left behind to keep their families together while those who they love are defending their right to do so.

In past wars the men were the ones who went off to war and the wives were left at home to wait for their man to return. Now men are not the only ones who go to war, fight and die for our freedom. Husbands are left behind as their wives go to war. They adapt to a lifestyle that may be strange to them at first but they endure just as their wives do in the fight for our freedom. The wives who are left behind endure for the same reasons. They, as do the men, work to keep the lifestyle that they were accustom to before separation by working hard to make a living and to keep the family together in a difficult time.

For the husband and wife couples who go off to war, the grandparents have taken up the job to raise their children until they return.

The pressures of struggle are many. Foreclosures on the veteran's homes are at an all-time high forcing some to make a major life style change and to make that change alone. To take care and to feed their families is the number one priority and at times this can be quite difficult and stressful.

If anything good has come out of the War on Terror it is that now there are organizations and special interest groups that help these families in need. These organizations and groups help all that need help and will continue to do so as long as there is one person or one family out there that may require assistance.

These husbands and wives of our soldiers are the true American unheralded heroes for what they endure in life that is just as difficult and lonely as their soldiers who fight thousands of miles away from home. They wait for their loved one to return. If their loved one does not return because they made the ultimate sacrifice for our freedom, then they will endure as before because they are the true American unheralded heroes of this great nation.

The Author's Perspective

Does an action or situation classify a person a hero? Wouldn't something that happened sometime in every one's life make them a hero? Are heroes made or does it just happen?

Were the US Astronauts who landed on the moon heroes for what they accomplished or because they landed there before the Russians did? Is Charles Lindberg a hero because of his transatlantic flight in 1927 that earned him the Medal of Honor and the title? Lindberg was also known as the first out-spoken anti-war advocate against the USA's involvement in World War II. His anti-war advocacy did little to hinder his hero status.

What constitutes a hero? Isn't everyone a hero in their own right? The next time you look into the mirror think about what you have done in your lifetime. Have you helped your fellow man? If you can answer yes, then the person looking back at you is definitely a hero. Now, saying all this, I am hungry so I'm going to go make me a hero sandwich!

Information References

www.carter-house.org
www.abrxis.com
www.genealogyforum.com
www.historynet.com
www.galenpress.com
www.hubpages.com
www.USAWWI.com
www.choctawnation.com
www.anzacday.org
www.wikipedia.org
www.cix.co.uk.com
www.us14thky.blogspot.com
www.msnbc.com
www.nppa.org
www.smithsonianmag.com
www.wral.com
www.olive-drab.com
www.vietnam.ttu.rdu
www.rIhtribune.com
www.almc.army.mil
www.cgsva.Ibbhost.com
www.vietnamgear.com
www.userpages.aug.com
www.kansaspress.ku.edu

Index

A

Acting Assistant Surgeon, 8
Aerial Gunner School, 30
Aerial photography, 26
Afghanistan, 83, 90, 125, 127
African Americans, 28
Agent Orange, 116
Airborne Warning And Control System, 121
Aircraft Carriers, 26, 64
AK-47, 125
Alexander, William, 3
American Civil War, 16, 19-20
American Military Advisors, 92
American Soldier, 29, 115, 128
An Khe, 95-96
An Nasariyah, 122
Anderson, Evelyn, 104
Andreotta, Specialist Glenn, 110
Ann Hester, Leigh, 125
Annapolis, Maryland, 11
APC, 122
Arab-Israeli War, 91
Arc of Triumph, 31
Arlington Cemetery, 12
Armed Personnel Carrier, 122
Armistice of WWI, 28
Army flight surgeon, 126
Army l-19, 94
Army Medal of Honor, 6
Army Republic Vietnam, 94
Artillery, 10, 19-20, 26, 40-41, 59, 68, 74, 81, 94, 123
ARVN, 94, 112
Atlanta, 30
Atlantic, 22, 30, 35, 47, 63, 87

Auld, David, 9
AWACS, 121

B
Baghdad, 120
ban me thuot, 103-104
Battle of Antietam, 10, 17
Battle of Chickamauga, 11
Battle of Cowpens, 2
Battle of Gaines Mill, 18
Battle of Gettysburg, 5
Battle of La Drang, 108-109
Battle of Malvern Hill, 18
Belleau Wood, 28
Binge, Michael, 104
Bird Dog, 74, 93-95
Black Death, 29
Bolt Action Rifles, 26
Bowser, Mary Elizabeth, 15
Bradley, 122
Bradley Fighting Vehicles, 122
Brady, Mathew B., 16
Brigade, 13, 91, 95
Brigadier General, 12, 73-74
British soldiers, 2
Brooklyn 13th Regiment, 11
Brooklyn's Greenwood Cemetery, 11
Brooks, Stephen, 111
Brown, Monica Lin, 125
Browning Automatic Rifle, 27
Bullard, Eugene Jacques, 30
Burr, Aaron, 3
Bush, President George, 109

C

C Co. 1st Battalion 20th Infantry, 109
Cairo, 21
Calley, William, 109
Cambodia, 91, 115
Cambodian border, 107
Camp Jackson, 29
Carondelet, 21
Carter, 6, 8-9
Carter House, 9
Carter, Jimmy, 6
Castro, Fidel, 91
Catlettsburg, 13
CBS News, 90-91
Central powers, 27
Cessna, 74, 93
Cessna 305, 93
Cessna O-1 Bird Dog, 74, 93
Chapelle, Dickey, 91-92
Charleston, 18
Charlie Company, 112
Charlotte Culver, 13
Cheney, Vice President Dick, 125-126
Chief Warrant Officer, 107, 126
China Sea, 102
Chinese Nung, 115
Chinook, 124
Chu Chi, 97, 103
Chu Lai, 92
Cincinnati, 21
City Class Ironclads, 20
Civil War, 3, 5, 8-10, 12-20, 22, 24, 50
Civil War Veteran, 12
Civilian Service, 75, 100
Clem, Johnny, 9, 11
Clinton, President Bill, 30, 53

Cluster Bombs, 120
Coast Guardsmen, 98
Cody, Buffalo Bill, 6
Colburn, Lawrence, 110
Colburn, Specialist Lawrence, 110
Cold War, 122
Colonists, 1-2
Columbia, South Carolina, 29
Columbus, Georgia, 30
Confederate, 9, 11-12, 14-16, 18, 22
Confederate Cavalry, 12
Confederate Colonel, 12
Confederate Intelligence Services, 14
Confederate Whitehouse, 15
Congress, 13, 124
Connecticut Regiment, 11
Cook, George S., 18
Cook, Johnny, 9-10
Cornum, Major Rhonda, 126
Crandall, Bruce P., 108
Croix De Guerre, 28-30, 40, 54, 73
Cumberland Gap, 13

D
Daimler, 27
Daisy Cutters, 120
Damita, Lili, 90
Davis, Benjamin O., Jr., 31
Davis, Jefferson, 16
de Gaulle, President Charles, 31
Decauville, 24
Department of Defense, 94, 101, 116
Desert Shield, 124, 127
Desert Storm, 119-120, 122-124, 126-127
Detroit, 10
Dien Bien Phu, 103

Distinguished Service Cross, 29-30, 38, 108
Do Ba, 112
Dustoff, 105, 107

E
Eads, James B., 20
Enfield Rifle, 24
Es, Hugh Van, 92
Europeans, 23

F
Father Nguyen Loc Hoa, 92
Field telephones, 26
First Battle of Bull Run, 16
Flame throwers, 27
Flynn, Errol, 90
Flynn, Sean Leslie, 90
Fort Sumter, 10, 18
Forward Operating Base Normandy, 126
France, 24, 28-31, 39-40, 47-48, 54-56, 68, 73
Franklin, Tennessee, 8-9
Freeman, Ed "Too Tall", 108
Freeman, Major Ed, 108
French Foreign Legion, 30
French General Goybet, 29
French Indochina War, 95
French Intelligence, 31
Ft. Wayne, 10

G
Gardner, Alexander, 17
Garros, Roland, 25
Gas masks, 26
Geneva Accords, 114
German, 24-25, 28-29, 31-33, 35, 41, 46-47, 56, 58, 61-62, 67-69, 74-75, 79, 103

Gibson, James F., 18
Global Positioning System, 120
GPS, 120
Grant, President Ulysses S., 12
Green Beret, 53, 115
Greenup, Kentucky, 12-14
Gun trucks, 95-96, 121-122

H

Hamilton, Alexander, 3
Hanoi Hilton, 103
Harlem Hell Fighters, 28
Henderson, Robert Henry, 9-10
Highway One, 91
Hill, Lori, 126
Hmong Army, 114
Hmong Soldiers, 114
Ho Chi Minh Trail, 104, 114
Hope, Bob, 102
House Armed Services Committee, 113
Howitzer Artillery Pieces, 26
Humvees, 125
Hungarian Revolution, 91
Hussein, Saddam, 119
Hydrophones, 26

I

I Corps, 92
Illinois Volunteers, 10
In Mermory Honor Roll, 117
Inventions Committee, 27
Iran, 90
Iraq, 83, 90, 95, 121-122, 124, 127-128
Iraqi Freedom, 126
Ironclad, 20, 22
Israel, 91

J

Jackson County Rifles, 10
Jewish Welfare Board, 100
Johnson, Henry Lincoln, 29
Johnson, Herman A., 30
Johnson, Specialist Shoshana, 127
Johnson, Willie, 9, 11
Junker, Hugo, 25

K

Kengkok, Laos, 104
Kennedy Administration, 114
Kentucky Infantry, 13
Khmer Rouge, 91
Khost, 125
Kinnear, Greg, 108
Kiowa Warrior Helicopter, 126
Korean War, 19, 45, 51, 74, 77, 79-80, 83, 85-87, 93-94, 104, 109
Kosin, Beatrice, 104
Kuwait, 122

L

L-19a Bird Dog, 93
Lafayette Flying Corps, 30
Land-ship Committee, 27
Land-ships, 27
Landing Zone X-ray, 108
Lao Communists, 114
Laser Guided Bombs, 120
Lee, Henry, 3
Lee, Robert E., 3, 15
Leprosarium, 103-104
Light Automatic Weapons, 26
Light Horse Harry, 3
Lincoln, Abraham, 6, 17
Little Willie, 27

London, 30, 56
Long Haired Warriors, 103
Lotz family, 9
Lotz house, 9
Lynch, Jessica, 126

M
M4 Rifle, 125
M1917 Enfield Rifle, 24
Maine, 5-6
Mash, 80-83, 87, 99
Massachusetts, 11
Mast, 106
Mcclellan, General George, 17
Mckenzie, Clarence, 9, 11
Medal of Honor, 5-6, 8, 10-11, 14, 24, 107-109, 116, 131
Medevac Pilots, 107
Medevacs, 104-105
Medical Services Command, 106
Medical Specialist, 100
Medina, Captain Ernest, 110
Merrimack, 22
Michigan Infantry, 10
Miles, Nelson, 6
Military Academy, 12
Minutemen Militia, 2
Mississippi, 14, 20-21
Mississippi River, 20
Mitchell, David, 110
Monsoon season, 104
Montagnard, 115
Morgan, Daniel, 2
Morgan, General George, 13
Mound City, 21
Murfreesboro, 10
My Lai, 109-110, 112-113

My Lai 4, 110
My Lai Massacre, 109, 112

N
National Catholic Community Services, 101
National Guard, 29, 74, 123
Native American Code Talkers, 32
Negroes, 15
Nein, Timothy, 125
New York City, 16, 18
New York Infantry, 29
New York National Guard, 29
New Zealand, 97
Newark, Ohio, 11
Nixon, President Richard, 107
North Vietnamese, 89, 91-92, 95, 114-115
North Vietnamese Army, 95, 115
Northerners, 15
Novosel, Michael J., 107
NVA, 95-96, 106

O
Oh-23 Raven, 110
Ohio Volunteers, 11
Okuley, Bert, 92
Olsen, Betty Ann, 104
Operation Black Ferret, 92
Operation Desert Shield, 124, 127
Operation Desert Storm, 119, 126
Operation Iraqi Freedom, 126
Operation Market Time, 98

P
Paktia Province, 125
Parachutes, 26
Paris, 30-31, 54, 89

Parisian, 30
Pathet Lao, 114
Patriot Missile, 121
Peers, Lieutenant General William, 113
Peninsula Campaign, 18
Pennington, Elizabeth, 13
Persian Gulf Wars, 119
Photocopiers, 121
Pittsburgh, 21
Pork Chop Hill, 109
Pows, 51, 104, 115, 124
Precision Guided Munitions, 119-120
Presidential Unit Citation, 116
Purple Heart, 29-30, 66, 68, 108

Q
Quang Ngai, 92, 112
Quang Ngai Province, 92, 112

R
Rathbun-nealy, Sp4 Melissa, 126
Red Cross, 100
Red Hand Division, 29
Reed, Ralph E., 9
Revolutionary War, 1-2, 4, 24
Richmond, 15
Rivers, Mendel, 113
Rocket Propelled Grenade, 126
Roosevelt, President Franklin, 100
Route 19, 95
RPG, 126
RPK Machine Gun, 125

S
Saigon, 89, 92-93, 99, 101
Salvation Army, 101

Sams, 95
Schwinn, Monika, 103
Scobell, John, 14
Scotland, 17
Scud Missile, 121
Sea Swallows, 92
Secretary of War, 5
Seven Days Battle, 11
Sherman, General, 10
Silver Star, 68, 95, 98, 125
Smart Bombs, 119
Smith, Robert M., 18
Smokeless Gunpowder, 27
SOC, 51-52
Soldiers Aid Societies, 13
Somalia, 90, 122
Son Tay, 115
South Carolina, 15
South Vietnam, 92, 104
South Vietnamese Air Force, 94
Southeast Asia, 99
Southern Vietnam, 97
Special Forces Aviation Section, 107
Special Services, 100
St. Johnsbury, Vermont, 11
St. Louis, 21
Stanton, Edwin M., 5
Steel helmets, 26
Stone, Dana, 91
Submachine Guns, 26
SVAF, 94
Swallows of Death, 30

T

Tanks, 26-28, 35, 122
Task Force Barker, 110, 112

Tet Offensive, 91, 99, 104
Texas, 60, 73, 102
Thompson, Hugh Jr., 109-110, 113
Time Life, 90
Time Magazine, 90-91
Tinclads, 20-21
Travelers Aid, 101
Trench Railways, 26
Tu Do Street, 92
Tubman, Harriet, 15
Tunnel Rats, 97-98, 103
Tuskegee Airman, 30-31

U

U-boats, 26, 35, 46, 67, 75
U.S. Air Force, 59, 68, 73, 94
U.S. Army, 8, 38-39, 67-68, 71, 74, 78, 93, 108, 121
U.S. Coast Guard, 67, 98
U.S. Marines, 65, 126
U.S. Navy, 67, 98
UH-1 Huey, 109, 112
Underground Railroad, 15
Union Army, 10, 15
Union Brown Water Navy, 20
Union Navy, 20, 22
United Press International, 91-92
Unknown soldier, 31
USA, 131
USAF, 59, 68, 73, 94
USCG, 67, 98
USMC, 95
USN, 67, 98
USO Club, 100-102

V

V-100, 122
Van Lew, 15
Verdun, 30, 40
Vermont, 11
Vicksburg, 10
Victory Belles, 101
Viet Cong, 97, 99, 103, 105, 110, 121
Viet Minh, 95
Vietnam Veterans Memorial, 116
Vietnam War, 19, 51, 53, 63, 73, 77, 79, 83, 89-93, 95, 97-99, 103-105, 108, 114-116, 127
Vietnamese Communist Forces, 91
Vietti, Eleanor Ardel, 103

W

WACS, 74, 99
Walker, Mary Edwards, 6, 8
War Department, 6
War on Terror, 127-129
Washington D.C., 75
Washington Penitentiary, 17
Washington, George, 3
West Point, 12
Western Flotilla, 21
Whiskey Rebellion, 2
White Elephants, 106
Wingate, Harry L., 28
Wireless communications, 26
Women's Army Corps, 74, 99
World War I, 23-24, 27-28, 35, 37, 39-40, 78
World War II, 19, 31-32, 43, 49-51, 53-55, 57-60, 62-63, 65-66, 69, 71, 73, 77-79, 81, 83, 85-86, 91, 100, 109, 122, 125, 131

Y

YMCA, 100
York, Alvin, 23
YWCA, 100

www.ingramcontent.com/pod-product-compliance
Lightning Source LLC
Chambersburg PA
CBHW031133090426
42738CB00008B/1073